Energy, Environment, and Sustainability

This books series publishes cutting edge monographs and professional books focused on all aspects of energy and environmental sustainability, especially as it relates to energy concerns. The Series is published in partnership with the International Society for Energy, Environment, and Sustainability. The books in these series are editor or authored by top researchers and professional across the globe. The series aims at publishing state-of-the-art research and development in areas including, but not limited to:

- Renewable Energy
- Alternative Fuels
- Engines and Locomotives
- Combustion and Propulsion
- Fossil Fuels
- Carbon Capture
- Control and Automation for Energy
- Environmental Pollution
- Waste Management
- Transportation Sustainability

More information about this series at http://www.springer.com/series/15901

Shantanu Bhattacharya · Akhilendra Bhushan Gupta
Ankur Gupta · Ashok Pandey
Editors

Water Remediation

Springer

Editors
Shantanu Bhattacharya
Department of Mechanical Engineering
Indian Institute of Technology Kanpur
Kanpur, Uttar Pradesh
India

Ankur Gupta
School of Mechanical Sciences
Indian Institute of Technology Bhubaneswar
Bhubaneswar, Odisha
India

Akhilendra Bhushan Gupta
Department of Civil Engineering
Malaviya National Institute of Technology Jaipur
Jaipur, Rajasthan
India

Ashok Pandey
CSIR-Indian Institute of Toxicology Research
Lucknow
India

ISSN 2522-8366 ISSN 2522-8374 (electronic)
Energy, Environment, and Sustainability
ISBN 978-981-10-7550-6 ISBN 978-981-10-7551-3 (eBook)
https://doi.org/10.1007/978-981-10-7551-3

Library of Congress Control Number: 2017959918

Printed on acid-free paper

This Springer imprint is published by Springer Nature
The registered company is Springer Nature Singapore Pte Ltd.
The registered company address is: 152 Beach Road, #21-01/04 Gateway East, Singapore 189721, Singapore

Preface

Energy demand has been rising remarkably due to increasing population and urbanization. Global economy and society are significantly dependent on the energy availability because it touches every facet of human life and its activities. Transportation and power generation are major examples of energy. Without the transportation by millions of personalized and mass transport vehicles and availability of 24 × 7 power, human civilization would not have reached contemporary living standards. The first international conference on 'Sustainable Energy and Environmental Challenges' (SEEC-2017) was organized under the auspices of 'International Society for Energy and Environmental Sustainability' (ISEES) by the 'Center of Innovative and Applied Bioprocessing' (CIAB), Mohali, from February 26 to 28, 2017. ISEES was founded at IIT Kanpur in January 2014 with the aim of spreading knowledge in the fields of energy, environment, sustainability and combustion. The society's goal is to contribute to the development of clean, affordable and secure energy resources and a sustainable environment for the society and to spread knowledge in the above-mentioned areas and awareness about the environmental challenges, which the world is facing today. ISEES is involved in various activities such as conducting workshops, seminars and conferences in the domains of its interest. The society also recognizes the outstanding works done by the young scientists and engineers for their contributions in these fields by conferring them awards under various categories.

This conference provided a platform for discussions between eminent scientists and engineers from various countries including India, USA, South Korea, Norway, Malaysia and Australia. In this conference, eminent speakers from all over the world presented their views related to different aspects of energy, combustion, emissions and alternative energy resource for sustainable development and cleaner environment. The conference started with four mini-symposiums on very topical themes, which included (i) New Fuels and Advanced Engine Combustion, (ii) Sustainable Energy, (iii) Experimental and Numerical Combustion and (iv) Environmental Remediation and Rail Road Transport. The conference had 14 technical sessions on topics related to energy and environmental sustainability and a panel discussion on 'Challenges, Opportunities and Directions of Technical

Education & Research in the Area of Energy, Environment and Sustainability' to wrap up the three-day technical extravaganza. The conference included 2 plenary talks, 12 keynote talks, 42 invited talks from prominent scientists, 49 contributed talks and 120 posters. A total of 234 participants and speakers attended this three-day conference, which hosted Dr. V. K. Saraswat, Member, NITI Aayog, India, as a chief guest for the award ceremony of ISEES. This conference laid out the road map for technology development, opportunities and challenges in this technology domain. The technical sessions in the conference included Advances in IC Engines and Fuels; Conversion of Biomass to Biofuels; Combustion Processes; Renewable Energy: Prospects and Technologies; Waste to Wealth—Chemicals and Fuels; Energy Conversion Systems; Numerical Simulation of Combustion Processes; Alternate Fuels for IC Engines; Sprays and Heterogeneous Combustion of Coal/ Biomass; Biomass Conversion to Fuels & Chemicals—Thermochemical Processes; Utilization of Biofuels; and Environmental Protection and Health. All these topics are very relevant for the country and the world in the present context. The society is grateful to Prof. Ashok Pandey for organizing and hosting this conference, which led to germination of this series of monographs, which included 16 books related to different aspects of energy, environment and sustainability. This is the first time that such a voluminous and high-quality outcome has been achieved by any society in India from one conference.

The editors would like to express their sincere gratitude to the authors for submitting their work in a timely manner and revising it appropriately at short notice. We would like to express our special thanks to Prof. Akhilendra Bhushan Gupta, Prof. Ankur Gupta, Mr. Pankaj Singh Chauhan, Dr. Rishikant and Prof. Shantanu Bhattacharya who reviewed various chapters of this monograph and provided their valuable suggestions to improve the manuscripts. We gratefully acknowledge the support received from various funding agencies and organizations for successfully conducting of the first ISEES conference SEEC-2017, where these monographs germinated. These include Department of Science and Technology, Government of India (special thanks to Dr. Sanjay Bajpai); TSI, India (special thanks to Dr. Deepak Sharma); Tesscorn, India (special thanks to Sh. Satyanarayana); AVL, India; Horiba, India; Springer (special thanks to Swati Mehershi); CIAB (special thanks to Dr. Sangwan).

At this stage of technology development, environmental pollution has become the greatest threat to the mother nature with rising level of pollutants in air, water and soil specifically due to industrial growth and increasing population. A significant amount of contaminants are being discharged into the environment and major percentage of this is discharged into water every day, causing a grave danger to all the water inhabitants and dependent living entities. The industrial effluents are discharged into the water streams of different rivers, ponds and canals instead of strict legal enforcement and directives, and this forms a major polluting source for the existing water resources. The water is being used for different industrial purposes like cooling, processing, transportation and heat treatment. The water gets mixed with various harmful chemicals, organic and inorganic compounds, acids, etc., during these processes and forms a highly contaminated effluent which is directly discharged into water streams by industrial houses. The wastewater contaminates the groundwater and also the inland water resources and soil and affects in a major way human, animal and plant life on a large scale. This critical issue compels the researchers to use technical and scientific skills in the field of wastewater treatment/remediation.

The current monograph is intended to spread knowledge to readers regarding the issue related to water pollution in recent times. This book describes various physical, chemical and biological methods, which are being utilized to provide a complete remediation or remediation to an extent where wastewater could be discharged safely into inland waterbodies or could be possibly reused for agriculture or other purposes. The main objective of this monograph is to provide the complete explanation of all the reasons, difficulties and processes to counter the water crisis with their advantages as well as disadvantages.

Kanpur, India — Shantanu Bhattacharya
Jaipur, India — Akhilendra Bhushan Gupta
Bhubaneswar, India — Ankur Gupta
Mohali, India — Ashok Pandey

Contents

About the Editors

Prof. Shantanu Bhattacharya is currently a Professor in Department of Mechanical Engineering and also the Head of Design Program at the Indian Institute of Technology Kanpur, India. He received B.S. degree in Industrial and Production Engineering from the University of Delhi, New Delhi, India, in 1996, M.S. degree in Mechanical Engineering from Texas Tech University, Lubbock, TX, USA, in 2003 and Ph.D. degree in Bioengineering from the University of Missouri, Columbia, MO, USA, in 2006. He served as a Senior Engineer with Suzuki Motors Corporation from 1996 to 2002. He completed his postdoctoral research at the Birck Nanotechnology Center, Purdue University, West Lafayette, IN, USA. He has around 70 international journal publications in areas related to sensor systems, nanotechnology, nanomaterials, advanced functional materials, photocatalysis, etc. He serves on the editorial board of several journals. He has been awarded IEI Young Engineer Award in 2010, ISSS Young Scientist Award in 2013, NDRF Design Award from IEI in Mechanical Engineering in 2014, and he holds a fellow position of IEI from 2016. His research interests include design and development of microfluidics and MEMS platforms, nanotechnology, nanomaterials, advanced functional materials, visible spectrum photocatalysis, sensors for various engineering applications.

Prof. Dr. Akhilendra Bhushan Gupta is a Distinguished Professor in Department of Civil Engineering, MNIT, Jaipur. His major research interests are environmental engineering, environment and health, biological waste treatment, environmental modelling, bioprocess engineering, wastewater treatment. His research area focuses on various aspects of water and air pollution and their impact on human health-evolved pathophysiologies for fluoride, nitrate and aluminium toxicities and linkages between air pollution and human respiratory health. He has also developed low-cost technologies for removal of fluorides and nitrates from drinking water/wastewater. Advance bioprocesses for waste treatment, characterization of PM fractions in air and developing field kits for biological analysis are his current areas of research. He is also a Life Fellow of IE, India, and Life Member of Indian Desalination Association, ISTE, Indian Society for Water Resources. He has published more than 30 peer-reviewed research articles and supervised 7 Ph.D. students in the research area of environmental science.

Dr. Ankur Gupta is currently Faculty at Indian Institute of Technology Bhubaneswar, Odisha. He has done his Ph.D. from Indian Institute of Technology Kanpur. He got 'ISEES Young Scientist Award' in 2017. He also got selected for 'BRICS Young Scientist conclave' held at Hangzhou, China. He is a Lifetime Member of Society of Energy, Environmental Sustainability (SEEC). He has published various international papers in referred journals, conference papers along with a patent. His research interest is micro-/nanomanufacturing and nanotechnology. He has won first prize in Electron Microscopy Contest held in 2013 and 2014. His work has been presented in several international conferences which were held in the USA (Material Research Society (MRS), San Francisco, California), Indonesia, etc., as well as in some reputed institutes within the nation like IIT Bombay, IIT Delhi, IIT BHU, IIT Guwahati, etc.). He has also been selected as one of the research scholars from India to attend the workshop and deliver a talk at National Institute of Nanotechnology (NINT), University of Alberta, Canada, in the area of nanofabrication and characterization.

Prof. Ashok Pandey, ***DPhil, FBRS, FNASc, FIOBB, FISEES, FAMI*** is Eminent Scientist at the Center of Innovative and Applied Bioprocessing, Mohali, and former Chief Scientist and Head of Biotechnology Division at CSIR's National Institute for Interdisciplinary Science and Technology, Trivandrum. He is Adjunct Professor at MACFAST, Thiruvalla, Kerala, and Kalaslingam University, Krishnan Koil, Tamil Nadu. His major research interests are in the areas of microbial, enzyme and bioprocess technology, which span over various programs, including biomass to fuels and chemicals, probiotics and nutraceuticals, industrial enzymes, solid-state fermentation. He has authored more than 1150 publications/communications, which include 16 patents, more than 50 books, 140 book chapters, 423 original and review papers, with *h* index of 78 and ~ 25,000 citations (Google Scholar). He is the recipient of many national and international awards and fellowships, which include Fellow of Royal Society of Biology, UK; Academician of European Academy of Sciences and Arts, Germany; Fellow of International Society for Energy, Environment and Sustainability; Fellow of National Academy of Science, India; Fellow of the Biotech Research Society, India; Fellow of International Organization of Biotechnology and Bioengineering; Fellow of Association of Microbiologists of India; Thomson Scientific India Citation Laureate Award, USA; Lupin Visiting Fellowship, Visiting Professor in the University Blaise Pascal, France; Federal University of Parana, Brazil; EPFL, Switzerland; Best Scientific Work Achievement Award, Government of Cuba; UNESCO Professor; Raman Research Fellowship Award, CSIR; GBF, Germany; CNRS, France Fellowship; Young Scientist Award He obtained his honorary doctorate degree from Univesite Blaise Pascal, France. He was Chairman of the International Society of Food, Agriculture and Environment,

Finland (Food & Health) during 2003–2004. He is Founder President of the Biotech Research Society, India (www.brsi.in); International Coordinator of International Forum on Industrial Bioprocesses, France (www.ifibiop.org), Chairman of the International Society for Energy, Environment and Sustainability (www.isees.org) and Vice-President of All India Biotech Association (www.aibaonline.com). He is Editor-in-Chief of *Bioresource Technology,* Honorary Executive Advisors of *Journal of Water Sustainability* and *Journal of Energy and Environmental Sustainability,* Subject Editor of *Proceedings of National Academy of Sciences (India)* and Editorial Board Member of several international and Indian journals, and also Member of several national and international committees.

Part I
General

Chapter 1
Introduction to Water Remediation: Importance and Methods

Shantanu Bhattacharya, Akhilendra Bhushan Gupta, Ankur Gupta and Ashok Pandey

Abstract Water has always been one of the most essential entities for survival of living systems almost since the evolution of life form. It is pertinent to study water systems in respect of enhancing the quality of water as it can be a significant source affecting the living systems in a direct or indirect manner. Nowadays, increased level of water consumption and correspondingly high levels of pollution have generated a prominent need for managing the water quality by maintaining safe levels for the water to be used in specific applications. In this respect, water remediation methods have taken a forward thrust in order to increase the water quality of potable water as well as that of industrial grade water in order to prevent contamination of natural water resources due to the discharge of industrial effluents. Several methods of water remediation such as physical, chemical, electrochemical and biological are discussed and worked out worldwide. This monograph gives a review of the importance of water remediation, various methods that are employed to remediate and some automated techniques which have been breakthroughs in the field of water quality control.

Keywords Water remediation · Wastewater · Industrial effluents
Photocatalysis · Organic contaminants

S. Bhattacharya (✉)
Department of Mechanical Engineering, Indian Institute of Technology Kanpur,
Kanpur, Uttar Pradesh, India
e-mail: bhattacs@iitk.ac.in

A. B. Gupta
Department of Civil Engineering, MNIT, Jaipur, Jaipur, Rajasthan, India
e-mail: abgupta.ce@mnit.ac.in

A. Gupta
School of Mechanical Sciences, Indian Institute of Technology Bhubaneswar,
Bhubaneswar, Odisha, India
e-mail: ankurgupta@iitbbs.ac.in

A. Pandey
CSIR-Indian Institute of Toxicology Research, Lucknow, India
e-mail: ashokpandey1956@gmail.com

S. Bhattacharya et al. (eds.), *Water Remediation*, Energy, Environment,
and Sustainability, https://doi.org/10.1007/978-981-10-7551-3_1

Water is an essential part of every living entity, and it is required to fulfil different types of needs in every phase of the life cycle. Apart from drinking, water is used for cleaning, food processing, industrial operations and agricultural work. There is no substitute of water at present on the earth's crust. Although the water is available in plenty in oceans, such water is not useful for daily use. Only, 0.3% of total water present on the earth's crust is useful for the mankind. The other 99.7% is in oceans, soil, icebergs or floating in atmosphere in vaporized or aerosol form. Hence, the available water that is directly usable for sustenance should be wisely used and also should be protected from against contamination. Hence, it is the need and necessity of the day to rapidly find out ways and means to firstly conserve water and then also try and safeguard water and remediate water in case it is contaminated so that it can be reused to serve the nutrient transport needs of living organisms. Unfortunately, with an increase in population, industries and climate changes, the situation arisen poses a major challenge to usable water. Rising pollution in rivers and other water bodies due to effluent discharge from various industries is identified as a major source of contamination. Surface water can be easily contaminated in comparison with groundwater due to higher accessibility. Water which is contaminated can be remediated/retreated to some extent for reuse. Different types of contaminants can be removed from water by using traditional as well as modern approaches depending upon the requirements. In the traditional approach, the water can be filtered by passing through a column of *bed and bank material*. This type of filtration technique provides a great alternative to water supply and treatment. Apart from filtration, *riverbank filtration systems* also provide protection against organic matter, thus modulating the biochemical oxygen demand levels. Some of the common subsurface processes available naturally on the earth's crust such as *Leaching, hydrolysis, precipitation, oxidation and reduction* also help in the naturalized treatment of water. Similarly, for surface water remediation, *decentralized in-stream ecological wastewater treatment* approach can be deployed. These in-stream methods of treatment are perhaps cheaper involving an overall low maintenance cost. These methods provide a good self-purification capability naturally provided to water streams and other waterways providing naturally available good-quality potable water. The current monograph discusses all such aspects related to water and its remediation.

A very important environmental challenge that is posed by contamination of water is the presence of *emerging organic contaminants (EOCs),* which reduce the quality of drinking water drastically. Some of the products which are readily used by human beings such as *medicines, pharmaceutical products, pesticides, and industrial chemicals* contribute directly as emerging organic contaminants. Such EOCs are gaining a significant importance in today's water remediation scenario and are primarily involved in drawing the water technology road map primarily because they possess complex molecular nature and are very hard to detect and remove. Most of the *wastewater treatment plants (WWTP)* are not designed to remove such contaminants, and water technology is increasingly prompting the generation of domain knowledge in this new area and also prompting the creation of new technologies. There is always a huge possibility that these compounds and their metabolites may escape from the WWTPs and enter the aquatic environment through the treatment plant discharge and also through the sewage. Many of these chemicals are designated as health hazards, may have endocrine disruptive properties (EDCs) degenerating the hormone upregulation within living systems and may have deleterious consequences to the directly affected aquatic and terrestrial life form. Physico-chemical and biological properties govern the fate of EOCs during wastewater treatment, and subsequent discharge of the untreatable constituents may pose the health challenges discussed above. Therefore, it becomes essential to improve treatment technologies in context of ability to remove EOCs and thus pose overall less environmental risk. The use of physical, chemical and biological methods such as adsorption, ultrafiltration, electrodialysis, Fenton reaction, and photocatalysis has been extensively studied as new domain in this area, and such techniques are effective in removing some of the EOCs from wastewater although there are engineering requirements which are worked upon to make some of these techniques scalable with industrial effluent treatments.

Of significant mention is the *Photocatalysis* which is one of the most explored phenomena to obtain potable water in spite of organic contamination in groundwater reserves due to the leaching of fertilizer products and other organic species. The term *photocatalysis* is understood as a chemical reaction that is caused by photoabsorption behaviour of a material which is commonly known as a *photocatalyst*. Water molecules absorbed over the surface of photocatalysts in normal environments are used to participate in an electron transfer reaction which breaks down the molecules absorbed along with water over the surface of such catalyst. This carried out a direct breakdown of higher organic molecular species into lower molecular sizes, and such size reduction may enable selective absorbtion of these smaller molecules over suitable surfaces. The higher-sized molecules although adsorbed over such surfaces may quickly create stearic hindrance and may allow the diffusional migration process of other unabsorbed species to slow down. A suitable class of catalysts which are efficacious to perform photocatalysis are the semiconductor materials with reasonable band gap which may furnish electron transport through short levels of irradiation. Further, it is very much possible to tailor the band gap as a function of structure, and morphology is the structuring scale that is suitably shifted to the mesoscale. In this event, creation of high-density defect states within the mesostructures may allow the electron transport process to be very fast even with visible light irradiation. There are two distinct sources of energy extensively utilized in the photocatalytic applications in water treatment: solar energy and artificial ultraviolet (UV) light.

The monologue further describes how water splitting can help to generate a hydrogen economy which is the dream of several nations across the globe. Water splitting—more specifically known as electrolysis of water into oxygen and hydrogen if carried out in an efficient manner—would be a key technological component of a hydrogen economy. Reaching the goal of economical photoelectrochemical (PEC) water splitting may likely involve one to combine the physics of efficient solar absorbers with high-activity electrocatalysts for the hydrogen and oxygen evolution reactions (HER and OER). Water splitting processes coupled to energy storage systems can also be another way of converting waste to energy and may provide an indirect route to remediate the environmental challenges associated with wastewater.

Bioelectrochemical systems (BESs) also provide an energy-efficient pathway for wastewater treatment and have attracted the attention of many people in the past decade. A BES is a bioreactor that converts chemical energy from chemical bonds of the organic compounds into electrical energy through catalytic reactions that are brought in by microorganisms under anaerobic conditions. The recent energy crisis has reinvigorated a high level of interest in BESs among the research community as a way to generate electric power or hydrogen from biomass without a net carbon emission into the ecosystem. BESs can also be used in wastewater treatment facilities to break down organic matters and recover valuable products. Soluble organic matter such as monosaccharide, acetate derivatives, alcohols and others, present in wastewater, are oxidized by electrochemically active bacteria (EAB) in the anodic chambers, and electrons and protons are produced that may be stored and harnessed in future applications.

Drinking water needs to follow technical specifications with all the components allowed in some permissible limits. The halides are another class of species which make it necessary for filtration processes to exist. For fluorides, the permissible limit is <1.5 mg/L in drinking water samples which is a sacrosanct standard. If limit of fluoride increases beyond this limit, it imposes significant effects to teeth and bones like dental fluorosis and/or skeletal fluorosis. Besides fluorosis, excessive intake of fluoride also causes muscle fibre degeneration, low haemoglobin levels, thyroid, gastrointestinal problems, abdominal pain, reduced immunity, etc. The alternative solution to this problem can be either to use alternative water resources (like surface water, rainwater and low-fluoride groundwater) or dilution, but these solutions alone cannot cater the high demand of potable drinking water due to its insufficiency. So de-fluoridation of drinking water is the only practical option to overcome the problem of high fluoride in drinking water. Hence, it is very necessary to bring down the fluoride level to permissible limits from our water resources. The fluoride remediation can be basically carried out using three basic strategies, precipitation and coagulation-based remediation techniques, membrane-based filtration techniques (reverse osmosis, nano-filtration, dialysis and electrodialysis) and adsorption-based remediation techniques (using adsorbents like alumina/aluminium-based materials, clays and soils, calcium-based minerals, synthetic compounds and carbon-based materials). The new advancements in the fluoride remedial methods are also emerging and have been discussed in this monograph.

Large-scale water remediation studies are these days possible due to recent advancements in the field of remote sensing and geographic information systems (GIS). Remote sensing-based techniques have been found to assist with water quality monitoring and management; however, GIS is what makes it possible to seamlessly link water quality monitoring in space and time with remediation strategies at reduced costs. GIS has also facilitated the collection of data and communication of water quality hazards in real time, through display maps. Water quality parameters such as chlorophyll-a, algae bloom, turbidity, suspended sediments and mineral content in water bodies (both surface and groundwater) have been studied using improved spectral and spatial resolution sensors and geospatial modelling techniques at low cost and with greater accuracy. High-resolution mapping of contamination zones through some of these rapid and accurate monitoring techniques helps in allocating remediation efforts to the critically affected areas quickly.

The entry of nanotechnology into the wastewater treatment engineering exemplified a noteworthy advancement, and as such, nanoparticles (NPs) have been exhaustively studied for possible remediation applications. Although NPs have a challenge related to their low levels of identifiability as foreign bodies by living systems, they do possess a high level of adsorption kinetics necessitating their increased usage in water remediation. The wastewater treatment with dispersed NP suspensions is still a fledgling area for new knowledge generation and to some extent antagonistic on grounds related to health and environmental safety. NPs may pose challenges related to ecotoxicity which may be a retrograde in their overuse. Theoretically, porous nanocomposites containing nanoparticles trapped within nanopockets is architectures which people do exploit heavily to utilize both the best of surface area and less mobility of nanoparticles. A range of carbon nanocomposites containing NP of semiconductor photocatalysts has worked very well in the field of UV and visible-light-induced photocatalysis of wastewater. Various carbon-based nanocomposites which are commonly used for water remediation are carbon–ZnO, graphene–SiO_2/Cu_2O, graphdiyne–ZnO, carbon nanoparticles, gold and platinum nanoparticles, carbon nanotubes–TiO_2 nanotubes, carbon aerogel–TiO_2, vertically aligned multiwalled carbon nanotube array–TiO_2, graphene nanoplatelets–SiO_2, multiwalled carbon nanotubes, metal-doped ZnO nanohybrid, multiwalled carbon nanotubes–TiO_2–SiO_2, carbon nanofibres–Ag–TiO_2, carbon nanotube–Ag_3PO_4 in Pickering emulsions, carbon nitrogen-doped TiO_2–SiO_2, carbon–Ag–TiO_2, etc. Different nanocomposites have a common benefit of effectively degrading the organic effluents in that the application of photocatalysis is directed to efficient and whole degradation and formed simple products that are environmentally safe; however, partial mineralization could have resulted in the formation of toxic by-products.

Carbonaceous material-based adsorption is further widely adopted as a method for dye wastewater treatment. Dye is introduced through textile, leather, paper, plastic and cosmetics industry effluents. The commonly adopted methods such as adsorption using activated carbon and filtration cause generation of a large amount of sludge and solid waste leading to high treatment cost. In recent years, incorporation of microwave (MW) with the carbon compounds to attain compounds such as granular activated carbon has served as a next-gen technology in the adsorptive removal/degradation of dye from wastewater. Such an area also is very novel and as such explored in this monograph as a novel future direction to water remediation.

This research monograph presents both fundamental science and applied innovations on several keys and emerging technologies for water remediation. Specific topics that are covered in the current monograph include:

- Water pollution, human health and remediation
- Water remediation through natural filtration
- Remediation of contaminated urban streams: a decentralized ecological wastewater treatment approach
- Treatment technologies for emerging organic contaminants removal from wastewater
- Role of photocatalysis in water remediation
- Water splitting by using electrochemical properties of material
- New age of wastewater treatment employing bioelectrochemical systems
- Remediation of industrial effluents
- Recent advancement in bioaugmentation strategies for process industries wastewater (PIWW) treatment
- Fluoride remediation from drinking water
- Applications of remote sensing and GIS in water quality monitoring and remediation: a state of the art review
- Recent advances in carbon–semiconductor nanocomposites for water remediation
- Dye wastewater treatment using carbonaceous materials by microwave-assisted methods

The topics are organized into five different sections: (i) general, (ii) water remediation: basics and natural water remediation, (iii) various techniques of water remediation, (iv) use of carbon-based composites for water remediation with a hope that such an assembly of various domains of water remediation will help to generate awareness of the latest technology trends in the water technology industry.

Part II
Water Remediation: Basics and Natural Water Remediation

Chapter 2
Water Pollution, Human Health and Remediation

Komal Jayaswal, Veerendra Sahu and B. R. Gurjar

Abstract Water is considered as the most essential source of life and important part of our natural resources. Due to rapid industrialization, urbanization and lack of awareness among people to consider water as a crucial commodity, around 80% of the world population is now facing water supply and security threat. In fact, about 2% of planet water is fresh and only 0.036% of water is accessible for use, rest 1.96% is present in polar ice caps, underground wells and aquifers. Furthermore, freshwater resources are progressively becoming unavailable due to huge amount of pollution in drinking water sources and also due to ignorance by human or industries and government authorities. The increasing water contamination by discharging untreated effluent is major problem faced by humanity worldwide. For this, government authorities and other organizations concerned about water conservation, awareness among people, strict laws about water security, sustainable and cost-effective wastewater treatment technology to overcome water pollution and water shortage problem for humans and biodiversity.

Keywords Water pollution · Wastewater treatment · Bioremediation
Ecology

2.1 Introduction

Water is the most vital among all natural resources and considered as the most important for life. Due to rapid industrialization, urbanization and lack of awareness among people to consider water as a crucial commodity, around 80% of the world population is now facing problem related to water supply and security (Vorosmarty et al. 2010). While planet earth is mostly covered with water, however, only about 2% of water can be considered for use and rest 98% is seawater that is unsuitable

K. Jayaswal (✉) · V. Sahu · B. R. Gurjar
Department of Civil Engineering, Indian Institute of Technology Roorkee,
Roorkee-Haridwar Highway, Roorkee 247667, Uttarakhand, India
e-mail: jayaswal.komal@gmail.com

S. Bhattacharya et al. (eds.), *Water Remediation*, Energy, Environment,
and Sustainability, https://doi.org/10.1007/978-981-10-7551-3_2

for drinking due to presence of salts. In fact from 2% of fresh water only 0.036% is accessible for use, rest 1.96% is present in polar ice caps, underground wells and aquifers. Furthermore, freshwater resources are progressively becoming unavailable due to huge amount of pollution in drinking water resources due to ignorance by human or industrial and government authorities (Kumar Reddy and Lee 2012).

Due to ignorance a large number of water resources get contaminated and become unsuitable for drinking and other purpose. Moreover, remaining freshwater resources gets progressively polluted due to discharge of untreated pollutants, chemicals and hazardous waste from thousands of industry directly into water body. Evolving industrialization and rapid population growth lead to more water demand that definitely marks water as a precious and fundamental commodity (Vorosmarty et al. 2010).

Emergent pollutants that are entering in water supply systems are pesticides, synthetic fertilizers, chemical compounds like dyes, heavy metals, hormones, personal care products, detergents, pharmaceuticals product that directly or indirectly may enter into the aquatic system and finally affect human health. Apart from this, waterborne pathogenic microbes enter the water system from research laboratory, hospitals, untreated sewage, septic tanks and various tanneries, food processing and meat packaging industries (Schwarzenbach et al. 2010). Pollutants have direct effect on human health and ecosystem. Apart from anthropogenic sources, other natural factors that strongly affect water quality are floods, storm, volcanic eruptions, earthquakes, etc. Major water pollutants have been shown in Fig. 2.1. New techniques have to be developed in coming years, for clean water supply otherwise water scarcity may lead to water wars and insecurity at social and political levels.

Fig. 2.1 Sources of water contamination (Kumar Reddy and Lee 2012)

Without any proper strict regulation by government authorities and concerned organizations, this severity of problem cannot be controlled. Lawmakers must impose strict policy for industries, organization and people for proper waste management, treatment of discharge effluent, proper working of treatment plant and strictly follow the effluent discharge standard. The increasing water contamination by discharging untreated effluent is major problem faced by humanity worldwide. In fact, in developing country like India even though facilities are available but due to mishandling practices and lack of sanitation awareness, water resources are not properly utilized. In India, this problem is related to both rural and urban areas, for example in most of the Indian city collection of household drinking water is dependent on season and infrequent supply results in storage of water. During storage of water, there is a chance of contamination and level of contamination increases by number of factors like quality of container used, place of storage, water collection site and handling practices (Brick et al. 2004). To overcome this problem, government authorities and other organization have to issue strict laws about water regulation and security and also develop cost effective and sustainable wastewater treatment technology to overcome water pollution and water shortage problem for humans and biodiversity. Apart from this awareness campaigns must launch to make people aware about importance of water (Kumar Reddy and Lee 2012; Vorosmarty et al. 2010).

2.2 Background

In twenty-first century, major problems faced by human society are interrelated to water quality and quantity issues. Furthermore, in coming decades these problems are going to be intensified in wider sense due to climate change that will result in melting of glaciers, rise in water level and temperature and change in water cycle pattern that is going to increase more frequent floods and droughts. A warmer climate can hold more moisture, causing 7% increase in water vapour with rise in every degree centigrade of temperature. Water vapour acts as greenhouse gas that drives earth's hydrological cycle and change in climate (http://earthobservatory.nasa.gov/Features/Water/). Lack of proper sanitation has directly related to lack of safe drinking water which affects human health, for example, untreated excreta enters into the food chain causing severe health issues. Other major problems related to water security are contamination of water source by pathogenic microorganisms and toxic chemicals which further lead to bioaccumulation in aquatic water plants and other living organisms including seafood and fish. Pollutants from domestic, agricultural and industrial sources merge to freshwater sources like rivers, lakes and groundwater aquifers without proper treatment. These pollutants cause physical, chemical and biological pollution of natural water resource, and it has become public concern in all parts of the world (Schwarzenbach et al. 2010).

Nutrients and natural organic constituents like nitrogen and phosphorus are present in relatively small amounts such as in milligram per litre, generally considered as macropollutants. The primary source of these pollutants is from agriculture by excessive use of fertilizers and animal manure, from fossils fuel used in industry, power and transportation sector which increases amount of nitrogen in air. Other sources of nutrients are from sewer system. The effects of these pollutants are generally well known, but the major challenges are in developing a sustainable treatment system, as high amount of nutrients can lead to generation of biomass concentration, eutrophication in lake and oxygen depletion. (Lohse et al. 2009; Heisler et al. 2008). Due to high concentration of salt in water, it becomes unfit for drinking and also inhibits the growth of agricultural crops. This condition is present near coastal areas in India and China. Intrusion of excessive marine salt into groundwater near coastal area due to overexploitation of aquifers which lead to inhibition of agricultural crops and direct mixing in fresh water makes it unfit for use (Kaushal et al. 2005; Kumar et al. 2015; Schwarzenbach et al. 2006).

Sometimes, even small concentration of synthetic and natural pollutants may exert adverse effect when present in mixtures, and accumulation of these pollutants causes acute and chronic effects on human health. For example, total inorganic nitrogen total phosphorus discharged into river worldwide is 21 and 5.6 (10^6 tons per year), respectively. Similarly, around 0.3–1 (10^6 tons per year) heavy metals like zinc, chromium, nickel, mercury, cadmium, lead get discharged into the river (Schwarzenbach et al. 2006). Apart from macropollutants, acute health effect may also be generated due to pathogenic microbes present in polluted water, so to control the effect of micropollutants on human health and aquatic life, appropriate and sustainable water treatment technology must be applied for preventing pollutants to enter water supply system (Schwarzenbach et al. 2010). Different treatment system has been evolved, but to select an appropriate treatment system based on type of pollutant entering in water system, proper maintenance and efficient removal have to be considered. However, selection method often signifies a challenge not only in the sense of technology but also for economic, social and political viewpoints (Filippelli 2008). Local needs, preferences and practice of local people govern the social issues for planning and development of overall technology framework. Apart from this, well-framed law and clear jurisdiction owned by government body control the implementation of technology (Massoud et al. 2009).

Globally, about 30% of available freshwater is consumed by industries and municipal corporation (Cosgrove and Rijsberman 2000) that generate a large amount of wastewater containing pollutants with different concentration. In majority, wastewater is discharged in natural water system without treatment or undergoes treatment that does not effectively eradicate most of the pollutants. Other sources are also responsible for micropollutants, these are runoff of million tons of pesticide from agricultural land, other from oil and gas spills worldwide (Eliopoulou and Papanikolaou 2007) and also from human-driven toxic chemicals such as heavy metals, metalloids and alloy.

Additional natural micropollutants are incorporated from biological sources, which produce taste and odour in water, and these are not a major toxicological problem, but it leads to great aesthetical concern. There are also a large number of hazardous waste sites, for example abandoned industrial and previous military sites, from where toxic pollutants may find their way directly into natural water system, such as groundwater. In India, hazardous waste sites is present in states like West Bengal, Uttar Pradesh, Gujarat, Kerala, Madhya Pradesh, Orissa, Maharashtra and Karnataka, having heavy metal and aromatic hydrocarbon in high concentration (http://www.cpcb.nic.in/LIST_OF_HW_CONTAMINATED_SITES.pdf).

Report shows that around 100,000 toxic chemicals have been registered, most of them are in daily use (Schwarzman and Wilson 2009), so there are numerous passages by which these chemicals may enter freshwater system like industrial discharge directly into water body, toxic chemicals from agricultural fields and also from biological production of toxins and malodorous compounds (Schwarzenbach et al. 2006). In this chapter, we attempt to focus on worldwide water pollution problem, different types of natural and synthetic water pollutants present in natural water body and their sources, effects on human health, biodiversity and their remediation.

2.3 Agricultural Water Demand, Its Uses and Effects

A large number of chemicals are used to maintain crop production and to increase agricultural growth rate every year. Other than this, several tons of pesticides are consumed to control pests, weeds, insects and other microorganisms. These pesticides and other chemicals generally contain hundred different types of active ingredients that directly impart toxicity to humans and other biota. These chemicals not only pollute soil, water resource, biota and ecosystem but also it stretches its effect to farmers and consumers (Galt 2008; European Commission 1991), and these contamination due to their continuous exposure to living organisms are of great concern.

Very high concentrations of pesticides and their intermediate products such as triazines and chloroacetanilides are detected in US rivers, and due to their high permissible range, it can effect nontarget organisms in soils and whole aquatic systems and also affect the quality of surface and groundwater (Gilliom 2007). Pesticides like dichlorodiphenyltrichloroethane (DDT), hexachlorocyclohexane (HCH), aldrin and dieldrin which are organochlorine in nature are commonly used in developing countries like India due to low cost and effectiveness among several pests. Use of pesticides like HCH, DDT, endosulfan, phorate are common in some parts of India like Punjab, Maharashtra, Karnataka, Gujarat and Andhra Pradesh (Abhilash and Singh 2009). These pesticides are used for crops like cotton, vegetables, sugarcane rice.

Effective mitigation measures have to be designed to prevent pesticides and others chemicals to reach surface and groundwater. Pesticides runoff is dependent on hydrological properties of soil like water flow patterns, permeability, topography, meteorological conditions and absorption behaviour of chemical composition of pesticides (Leu et al. 2004). Pesticides applied to crop at first were degraded by soil microbes and chemical reactions, further it also absorbs in organic matter present in soil. It transforms into other metabolites but does not completely degrade. The final pathway of pesticides in environment takes place through volatilization in atmosphere, runoff in dissolved and particulate form to surface water bodies like river and mixing in groundwater through leaching process (National Research Council 1993).

Water contamination takes place through runoff of pesticides containing rainwater from roofs and roads in urban areas, which lead to pollution in drainage and sewer system. Pesticides application in agricultural practice causes direct exposure and health risk for workers, and extensive use of pesticides leads to water and soil pollution. In the area prone to high pesticides concentration, effective mitigation measures have to be applied by replacing or restricting extensive use of pesticides (Leu et al. 2004a; b). To maintain ecological balance and biodiversity, extensive use of pesticides in agricultural land has to be controlled and further direct exposure of pesticide chemicals to workers and farmers has to be regulated. Report shows that due to accidental exposure of pesticides, poisoning of 3 million people and 20,000 accidental deaths takes place per year in developing countries (United Nations Environment Programme 2007). The main role of the agricultural practices in most developing countries is to attain and sustain food security issues for the emerging populations, also keeping in mind about type of agricultural practices, pesticides and agrochemicals usage and their effect on human and other living being. With increasing trends in urbanization and industrialization, development in agricultural practices also increases which lead to water usage and quality issues. Due to extensive pesticides usage per hectare of agricultural land in developing countries, their concentration monitoring and effect assessment for human and plants health are often limited and implementation of regulations for maintaining pesticide concentrations are limited (Menezes and Heller 2008; Agrawal 1999).

2.4 Groundwater Contamination, Sources and Effects

2.4.1 Groundwater Contamination by Geogenic Sources

Aquifers geological composition generally causes leaching of toxic elements in water supply. Arsenic, fluoride, iron, chromium and selenium are main elements concerned with groundwater contamination. Arsenic among all these contaminants has caused extreme health effects and risk to the life of people in different countries. In Bangladesh, about 95% of the total population uses groundwater from 10 million tube wells. From this, around 60% underground water wells near Ganga and Brahmaputra river system contain arsenic beyond permissible limit as reported by World Health Organization (WHO) (Ahmed et al. 2004). High concentration of arsenic is due to weathering of arsenic-containing rocks in mountain ranges, deposition of organics in floodplains of rivers and due to high residence times of water in groundwater aquifers which cause anoxic condition for release of arsenic in water. Other factors that affect groundwater with arsenic contamination are mobilization of arsenic at high pH with oxygen-rich groundwater; this phenomenon occurs in arid areas such as US Midwest, eastern Australia and central Asia (Amini et al. 2008). Health effects caused due to high arsenic concentration are high blood pressure, neurological dysfunctions and pigmentation in limbs that lead to keratosis by accumulation in the skin, hair and nails (Chen et al. 2009). Chronic effect of arsenic leads to skin, lungs and internal organ cancer. It has been reported that carcinogenic effect of arsenic is caused due to contamination level greater than 50 μg As/L (Smith et al. 2002). Different technologies have been developed like membrane filtration, adsorption for treatment of arsenic contaminated water, but due to high cost of treatment, these methods cannot be used at places with alarming situations, so there is a need to develop cost effective, feasible and alternative techniques like rainwater harvesting, water sources from deep aquifers for rural areas in developing countries. For developing countries simple, effective and economical treatment technology is required. Furthermore, alternative methods like rainwater harvesting and adsorption techniques have to be implemented to improve water quality.

2.4.2 Groundwater Contamination by Anthropogenic Sources

Groundwater contamination is related to geological formation of aquifers, where pollutants enter into underground water table and pollutants transported to other stretch of drinking water source. Groundwater contamination results from municipal solid waste landfills sites, nuclear waste site, hazardous waste sites, accidental spills, runoff from agricultural land and waste discharge from industry. Many abandoned sites where illegal dumping and discharge of waste materials take place can result in groundwater pollution. Around 100 million tons of waste gets discarded that contains radioactive and hazardous water which enters groundwater table through leaching by contaminated landfills (United States Environmental Protection Agency 2008). Some of the contaminated sites are under maintenance and control conditions but most of them are expected to release toxic waste to environment. In addition, thousand tons of oil, petroleum and other chemical spills take place every year which resulted in pollution of water during storage and transportation process. Types of toxic chemicals that fluxes from contaminated site are fuel hydrocarbon, chlorinated ethane, pesticides, radio nuclear waste from nuclear weapon testing sites, explosive chemicals from ammunition plants, heterogeneous and non-categorized disposed waste and wastewater (Schwarzenbach et al. 2010). In fact, leachate composition at landfill sites is heterogeneous and complex as only few contaminant species are identified, so if there is need to monitor type of waste dispose at these sites, continuous monitoring of leachate material and groundwater is needed. Groundwater is directly used for drinking, and these persistence contaminants took several decades to decompose, so monitoring and assessment are necessary to check health risk and exposure of chemicals on human body. For this implementation of cost effective, reliable and sustainable approach is needed.

Groundwater contamination quantification and treatment approaches are phytoremediation, site excavation, pump-and-treat procedures and permeable reactive barriers (Salt et al. 1998). However, most of the remediation methods are very expensive and also more time is required for decay process. Treatment methods include removal of contamination source, catalytic reaction to immobilization of metals and conversion to biodegradable (organic) materials. Nowadays, remediation strategies include microbial or abiotic degradation processes. Presence of high amount of carbon source along with microbes at contaminated site leads to anoxic conditions, which further proliferates the growth of different microbial community at site which helps in degradation of certain types of waste materials. Anoxic condition helps in growth of metal reducing microbes which transforms organic and inorganic pollutants. Thus, for elimination of waste, proper assessment and understanding of type of contaminant along with its reactivity and mobility with other contaminants, geographical conditions of landfill site are required. Analytical tools such as stable isotopes analysis method provide information about specific compound, its transportation and transformation behaviour in groundwater during pollution and accidental spills conditions (Hofstetter et al. 2008).

2.5 Surface Water Contamination, Sources and Effects

Mining operations trigger in large amount of waste deposits which cause significant environmental problems. Generally, these generated wastes get oxidized by air and weathering by precipitation which further contaminates water resources. Mining activities result in large amount of waste to finally yield significant end products. Around 1,000 tons of waste materials are produced per kilogram of pure metal, during extraction of rare metals like gold, copper and nickel. Ores of iron, coal and copper contain large amount of sulphide, which further oxidizes in presence of air and water and forms sulphuric acid by acid mine drainage (Blodau 2006). It has been estimated that worldwide about 20,000 km of river stretch and 70,000 ha of lake and reservoir area are effected by acid mine effluent (Johnson and Hallberg 2005). Further, mining of precious metals uses large amount of synthetic chemicals, power supply and water that lead to environmental pollution and risks.

For example, during mining operation of gold, chemical extraction method is used that requires toxic elements like mercury and cyanide that directly affect human and environment. This chemical extraction process consists of mercury amalgamation and cyanide extraction process (Schwarzenbach et al. 2010). In above methods, gold traces are mixed with liquid mercury, which is further extracted by heating and evaporation in atmosphere. While heating process, this mercury gets released to atmosphere and local people and workers are directly exposed to neurotoxic metals. Biomagnification of mercury takes place during its discharge in surface water which causes health risk problem to aquatic system, organisms and human populations. On the other hand, at lower gold concentration and larger volume, cyanide method is used. In this method, oxidative leaching of gold takes place, which forms complex in aqueous solution, after that adsorption takes place and recycling of cyanide solution takes place. In general, for extraction of 1 kg of gold, around 700 tons of water and 140 kg of cyanide are required (kg) (Mudd 2007). Health issues related to cyanide are that it blocks enzyme function in respiratory chain process in higher organisms that cause sudden death. For adult human, 100 mg of cyanide is highly toxic while for fishes a much lower concentration that is 0.05 mg/L in water also causes death (Schwarzenbach et al. 2010).

Improved extraction processes should be implemented for gold extraction, for this toxic element like mercury and cyanide need to be replaced by less toxic extraction chemicals like halogens and thiourea (Dudka and Adriano 1997; Akcil 2003). Apart from this, upgraded sustainable processes and safety methods are needed to be adapted for extraction process. Enforcement of strict law should be there for health and environmental risks related to mining operations.

2.6 Pharmaceuticals and Personal Care Products in Wastewater and Drinking Water

Municipal wastewater significantly contributes to micropollutants in the form of organic pollutant into the environment; mostly consist of pharmaceutical and personal care product. Pharmaceuticals products include hormone regulator, sedative, contraceptives, lipid regulators, antibiotics, painkiller, anti-inflammatory, anti-depressants and others medicines. Every year, new pharmaceuticals products are launched in market. In fact, every year pharmaceuticals products contribute 8% of total research and development expenses. It can be extrapolated with this data that 300 new pharmaceuticals compounds hit market every year. The world-level market of pharmaceuticals compounds was 100,000 tons per year that were worth of US$773 billion, with the highest per capita sales in the USA with US$676 (Schwarzenbach et al. 2010; Kummerer 2013).

These products are highly bioactive and their concentrations are generally in nanograms to micrograms per litre of water and wastewater system. In municipal raw sewage, the range of pharmaceutical products is in micrograms per litre which confirms that sewer lines are main path for pharmaceutical discharge in water bodies. At present, pharmaceutical compounds are separated from wastewater system by biodegradation and absorption from sludge generated in system (Schwarzenbach et al. 2010). Due to formation of metabolites, complete mineralization of compounds does not take place during biodegradation process. Therefore, from ecotoxicological point of view, toxic effect of the compound is not only due to parent compound but also transformed metabolites need to be considered. As such in case of iopromide, during an iodinated X-ray contrast medium process, 12 metabolites were identified, that are equally toxic in nature as parent compound (Schulz et al. 2008). In pharmaceutical compounds, toxicological effects are not only due to parent compound but also due to their transformed metabolite in wastewater system.

Studies show that metabolites having more number of hydrophilic metabolites have less toxic effect as compared to more hydrophobic compounds present in their parent compounds. Moreover, the main issues related to pharmaceutical compounds are their ecotoxicological effect, as this water is used directly or indirectly for drinking and other purposes. Filtration of pharmaceutical compounds and their metabolites takes place through riverbank infiltration in the aquifers (Schwarzenbach et al. 2010). In previous study, it is shown that out of 19 antibiotic present in surface water, only one has been identified in the bank filtrate (Heberer et al. 2008). Some reports also suggested that complete removal of all potential pharmaceuticals compounds cannot be certain through riverbank filtration. Due to toxicological effect, pharmaceutical compounds and their transformed metabolites products are carefully tested. Moreover, from toxicological point of view, pharmaceutical compounds are carefully tested. Development of a new pharmaceutical compound requires proper details of drug kinetics, pharmacology and toxicology such as genotoxicity, carcinogenicity, reproductive and development toxicity. The final phase before launching of pharmaceutical products is clinical trial. Final phase is the clinical trial before launching (Schwarzenbach et al. 2010). On the basis of above context, risk of toxicity to the consumers seems quite low. However, studies are conducted for evaluation of long-term effect of these products on human and environmental health. As wastewater is the main source for pharmaceuticals, so different wastewater technology has been evolved, such as ozonation, polishing pond and activated carbon. Ozonation process has high treatment efficiency, and it has feasible operation and maintenance cost (Hollender et al. 2009). Other process such as polishing treatment of effluent helps in treatment of endocrine disrupting compounds. Alternatively, activated carbon in granular or powdered form along with oxidation and nanofiltration and reverse osmosis can be used as a promising technique for removal of these compounds (Schwarzenbach et al. 2010).

2.7 Contamination by Waterborne Viruses and Microbial Pathogens

Problems related to sanitation, hygiene and waterborne pathogen are of the major concern in developing countries. As in developed countries, maintenance of water supply infrastructure related to sanitation is already installed. Unfortunately, in developing countries where large amount of sewage is discharged into water body without treatment, so improvement in basic sanitation facility and treatment technology is necessary; moreover, access to safe drinking water is the primary requirement. It is the current estimation that 67% of the population in developing countries may not be able to have a proper sewerage system till 2030 (Schwarzenbach et al. 2010). Currently, in developing countries, there is an imbalance between urban and rural areas for proper sanitation and access to safe drinking water. Report shows that nearly 1.1 billion people across the world lacks potable water and 2.6 billion people have no proper sanitation facility.

Lack of proper sanitation and limited access to safe water causes 1.6 million deaths per year worldwide and from this most of cases are of developing countries, and the main cause of disease related to drinking water in developing countries is pathogenic viruses, bacteria, protozoa and insects developing on contaminated water. One of the reports predicted that contaminated water is responsible for 15–30% of gastrointestinal diseases. Other diseases caused by consumption of unsafe water are diarrhoea, cholera, stomach infection, skin problem, nausea, typhoid fever and legionellosis. Most increasing waterborne disease is typhoid fever that is caused by Salmonella typhi and Salmonella paratyphi, respectively. Even acute viral diseases like hepatitis A and E, rotaviruses and the other protozoa-related diseases caused by Giardia lamblia are often found related to inadequate supply of safe water and sanitation practices (Schwarzenbach et al. 2010). Pathogens like Cryptosporidium parvum, Campylobacter jejuni, enterotoxigenic and enteropathogenic *Escherichia coli*, *Shigella* spp., or *Vibrio cholera* cause most of the chronic health diseases (Albert et al. 1999). Pathogenic microorganisms such as *E. coli* and cryptosporidiosis generally cause outbreak of diseases in developed countries, due to microbial distribution in warm water supplies and air condition from houses, complexes and hospitals.

Due to discharge of untreated sewage in water supply system or other natural water body, nowadays, health-related problems are linked to both wastewater and drinking water supply. Moreover, issues related to sanitation, proper access to safe water, wastewater treatment and enforcement of strict regulation for water safety and conservation are given low priority and low number of funds is allocated for this purpose. Improved sanitation, proper and safe water supply, linked sewerage line to every house, proper treatment and management of waste can help to eradicate these health-related issues. World health organization (WHO) along with the United Nations Children's Fund has launched as a millennium development goal (MDG) to help people with access to safe drinking water and basic sanitation facility by 2015 (World Health Organization UNICEF 2006).

2.8 Wastewater Treatment Technology for Municipal Sewage: A Self-sustainable Option

To improve sanitation and to maintain human health and ecosystem, it is necessary to develop a sustainable and effective wastewater treatment system. The main aim of municipal wastewater treatment is to eliminate organic matter in the form of biological oxygen demand (BOD) and chemical oxygen demand (COD), and nutrients such as nitrogen and phosphorus and pathogenic microorganisms. Removal of nutrients from wastewater effluent causes decrease in eutrophication phenomenon that leads to proper growth of aquatic organization and high dissolved oxygen (DO) in water. It has been estimated that in developed countries, 50–95% of wastewater gets treated while in developing countries about 80% of effluents is discharged in river and lakes without prior treatment (Kumar Reddy and Lee 2012). Wastewater from industries not only leads to organic pollutant but also it is a source of heavy metals, synthetic dyes and organic compounds.

Treatment processes such as advanced oxidation, membrane bioreactor, neutralization of acid and base in effluent, resource recovery within system and activated carbon processes help to reduce these pollutants very efficiently, but unfortunately, these processes are not implemented in developing countries. Now it's high time to develop effective mitigation measure for wastewater treatment, as due to increase in water scarcity, treated wastewater is not only used in agriculture and recreational work but also for drinking purpose. For example, wastewater is recycled in Windhoek, Namibia since 1973 for obtaining drinking water (Schwarzenbach et al. 2010). In any country for selection of sewage treatment system, economy is first priority. Many sewage treatment systems such as activated sludge process (ASP), membrane bioreactor (MBR) are very efficient for organic and nutrient removal, but due to high cost of construction, operation and maintenance either it is not preferred or are not being maintained properly due to lack of understanding of system. For developing country like India, economical aspects are a major thing to be considered for development of sewage treatment plant.

Apart from this in developing countries, sewage treatment and other environmental issues are of low preference in comparison to other developmental issues such as food availability, health, education and industrial development as it does not have direct economical significance. So installation of highly advance and costly treatment technology is not preferable for developing countries. Now it is a time to develop sustainable, cost effective and appropriate sewage treatment system which is techno-economically feasible. Sometimes for developing countries, it is very tough to define about appropriate treatment technology as there is a serious concern about selection, because low cost is not only the criteria but removal efficiency, low operation and maintenance, low sludge withdrawal of the system are the criteria that are to be considered. So under these circumstances, a combination of anaerobic-aerobic system has to be considered as appropriate wastewater treatment system for developing countries.

In this context, a combination of up-flow anaerobic sludge blanket (UASB) reactor and down-flow hanging sponge (DHS) reactor is considered as an appropriate sewage treatment technology for developing countries (Tandukar et al. 2005, 2006). Combination of anaerobic–aerobic system for wastewater not only provides better removal efficiency but also can provide high removal efficiency at minimum operation control.

DHS reactor was developed as a novel, effective and low-cost post-treatment for UASB effluent to meet desired effluent standard. The principle of this system is based on conventional trickling filter process, in which media is replaced with polyurethane for high sludge retention in the reactor. This high sludge retention is due to the presence of more than 90% of the void space resulting in high biomass entrapment in packing media, which finally leads to long solid retention time (SRT). Moreover, DHS rector generates less sludge due to longer SRT which provides enough time for mineralization of sludge in system itself. Due to presence of natural ventilation, there is no requirement of external aeration or any other energy source. Oxygen present in the atmosphere gets incorporate to wastewater, while effluent moves downwards in the reactor and finally gives high DO in resultant effluent (Tandukar et al. 2005). UASB–DHS system not only provides better removal efficiency in terms of organic matter and nitrogen removal but removes pathogenic microorganisms.

2.9 UASB-DHS System as a Novel, Cost-Effective Treatment Technology for Municipal Wastewater

Various studies have been conducted by using different generation of DHS sponge media on bench and pilot scale reactor. These reactors have shown an excellent ability to polish the quality of UASB effluent in terms of organic matter, ammonium nitrogen, and suspended solids and pathogens removal to meet the desired effluent standards. Full-scale study on DHS system was conducted in India, and experimental results showed that DHS reactor is efficient and can be taken as viable option as post-treatment method for UASB effluent (Okubo et al. 2015).

Water quality parameter of 15 different UASB-FPU-based STPs in India were investigated and analyzed and found out that this system is not efficient to remove organic matter, ammonium nitrogen and pathogens to meet desired Indian effluent discharge standards. So to understand the practical approach of DHS as a post treatment of UASB water quality profiles and retained sludge need to be investigated (Sato et al. 2006). Organic matter removal takes place in DHS reactor due to anaerobic and aerobic biological degradation and also due to adsorption in sponge media. At first, organic removal takes place due to proliferation of heterotrophic bacteria as dominant-microorganisms. After removal of organic matter, ammonia nitrogen removal takes place by autotrophic bacteria. Maximum removal of NH_3–N takes place in the lower half of reactor due to presence of nitrifiers that convert NH_3–N to NO_3–N in the presence of oxygen. Higher biomass concentration in sponge media results in better organic and nitrogen removal. Apart from this, pathogenic microorganism gets removed due to adsorption in sponge media, predation by higher form of microorganisms, competition of space and food allows growth of certain microorganisms. High amount of active biomass retained in the sponge of DHS and corresponding longer SRT ensures a high degree of treatment at minimum operational control. These properties are important to hedge against any hydraulic or organic overload to the system during the real operation as well as to reduce sludge production (Mahmoud et al. 2009). The performance of reactor was attributed to its distinctive characteristics like high DO uptake, dense sludge retention and adequate sludge activity in sponge media. The results verified the practical-scale applicability of DHS reactor to developing countries for effective treatment.

2.10 Conclusions

To overcome global water pollution problems, it is required to develop an effective approach that fits at technological, sustainable, economical and within policies limit to meet desired treatment standard. Hazardous compounds, volatile chemicals, synthetic and toxic compounds should not be allowed to enter food chain or other sources of human and ecological biodiversity area or treated onsite in a closed system. Human food production areas like agricultural sector require protection from compounds that have bioaccumulation property. Globally, agricultural sector faces problem related to higher production yield and water shortage, and at the same time, environmental safeguards and contamination in food chain should be secured. Geogenic and anthropogenic sources must be identified, and different approaches should be developed to prevent water sources from these contaminants. To provide proper sanitation and to fulfil water demand, efforts have to be taken to develop sustainable and cost-effective wastewater treatment system. Household-centred sanitation system needs to be developed for proper sanitation facility. To meet water crises and cleaning, large-scale water pollution UASB-DHS system can be considered as a novel approach for wastewater treatment.

2.11 Way Forward

- In spite of advanced wastewater treatment process, efforts have to be designed to reduce mixing of toxic and hazardous chemicals into aquatic water body. This can be accomplished either by onsite treatment of these waste before discharge or to substitute these chemicals with less harmful chemicals.
- Industry should emphasize on designing new chemicals that have less environmental effect. In addition to this, improved and advance wastewater treatment systems should be introduced.

- Mining activities lead to surface and groundwater pollution, so proper mitigation approach should be developed to improve and develop effective removal techniques.
- Decentralized wastewater treatment system should be preferred over high-cost centralized system to allow reuse of water and other nutrients locally.
- Development of fast, cost effective and reliable detection technique for micropollutants and pathogens should be developed for wastewater and water quality assessment.
- Interlinking rivers is one of the approaches to overcome freshwater demand in future.

References

Abhilash PC, Singh N (2009) Pesticide use and application: an Indian scenario. J Hazard Mater 165(1):1–12

Agrawal GD (1999) Diffuse agricultural water pollution in India. Water Sci Technol 39(3):33–47

Ahmed KM, Bhattacharya P, Hasan MA, Akhter SH, Alam SM, Bhuyian MH, Sracek O (2004) Arsenic enrichment in groundwater of the alluvial aquifers in Bangladesh: an overview. Appl Geochem 19(2):181–200

Akcil A (2003) Destruction of cyanide in gold mill effluents: biological versus chemical treatments. Biotechnol Adv 21(6):501–511

Albert MJ, Faruque ASG, Faruque SM, Sack RB, Mahalanabis D (1999) Case-control study of enteropathogens associated with childhood diarrhea in Dhaka, Bangladesh. J Clin Microbiol 37(11):3458–3464

Amini M, Abbaspour KC, Berg M, Winkel L, Hug SJ, Hoehn E, Johnson CA (2008) Statistical modeling of global geogenic arsenic contamination in groundwater. Environ Sci Technol 42(10):3669–3675

Blodau C (2006) A review of acidity generation and consumption in acidic coal mine lakes and their watersheds. Sci Total Environ 369(1):307–332

Brick T, Primrose B, Chandrasekhar R, Roy S, Muliyil J, Kang G (2004) Water contamination in urban south India: household storage practices and their implications for water safety and enteric infections. Int J Hyg Environ Health 207(5):473–480

Chen Y, Parvez F, Gamble M, Islam T, Ahmed A, Argos M, Ahsan H (2009) Arsenic exposure at low-to-moderate levels and skin lesions, arsenic metabolism, neurological functions, and biomarkers for respiratory and cardiovascular diseases: review of recent findings from the health effects of arsenic longitudinal study (HEALS) in Bangladesh. Toxicol Appl Pharmacol 239(2):184–192

Cosgrove WJ, Rijsberman FR (2000) World water vision: making water everybody's business. World Water Council, London

Dudka S, Adriano DC (1997) Environmental impacts of metal ore mining and processing: a review. J Environ Qual 26(3):590–602

Eliopoulou E, Papanikolaou A (2007) Casualty analysis of large tankers. J Mar Sci Technol 12(4):240–250

European Commission (1991) Plant protection products Directive 91/414/EEC. http://ec.europa.eu/food/plant/protection/evaluation/dir91-414eec_en.htm

Filippelli GM (2008) The global phosphorus cycle: past, present, and future. Elements 4(2):89–95

Galt RE (2008) Beyond the circle of poison: significant shifts in the global pesticide complex, 1976–2008. Glob Environ Change 18(4):786–799

Gilliom RJ (2007) Pesticides in U.S. streams and groundwater. Environ Sci Technol 41: 3407–3413

Heberer T, Massmann G, Fanck B, Taute T, Dünnbier U (2008) Behaviour and redox sensitivity of antimicrobial residues during bank filtration. Chemosphere 73(4):451–460

Heisler J, Glibert PM, Burkholder JM, Anderson DM, Cochlan W, Dennison WC, Lewitus A (2008) Eutrophication and harmful algal blooms: a scientific consensus. Harmful Algae 8(1): 3–13

Hofstetter TB, Schwarzenbach RP, Bernasconi SM (2008) Assessing transformation processes of organic compounds using stable isotope fractionation

Hollender J, Zimmermann SG, Koepke S, Krauss M, McArdell CS, Ort C, Siegrist H (2009) Elimination of organic micropollutants in a municipal wastewater treatment plant upgraded with a full-scale post-ozonation followed by sand filtration. Environ Sci Technol 43(20): 7862–7869

http://earthobservatory.nasa.gov/Features/Water/

http://www.cpcb.nic.in/LIST_OF_HW_CONTAMINATED_SITES.pdf

Johnson DB, Hallberg KB (2005) Acid mine drainage remediation options: a review. Sci Total Environ 338(1):3–14

Kaushal SS, Groffman PM, Likens GE, Belt KT, Stack WP, Kelly VR, Fisher GT (2005) Increased salinization of fresh water in the northeastern United States. Proc Natl Acad Sci USA 102(38):13517–13520

Kumar Reddy DH, Lee SM (2012) Water pollution and treatment technologies. J Environ Anal Toxicol 2:e103

Kumar KA, Priju CP, Prasad NN (2015) Study on saline water intrusion into the shallow coastal aquifers of Periyar River Basin, Kerala using hydrochemical and electrical resistivity methods. Aquat Procedia 4:32–40

Kummerer K (2013) Emissions from medical care units. In: Pharmaceuticals in the environment: sources, fate, effects and risks, p 27

Leu C, Singer H, Stamm C, Müller SR, Schwarzenbach RP (2004a) Variability of herbicide losses from 13 fields to surface water within a small catchment after a controlled herbicide application. Environ Sci Technol 38(14):3835–3841

Leu C, Singer H, Stamm C, Müller SR, Schwarzenbach RP (2004b) Simultaneous assessment of sources, processes, and factors influencing herbicide losses to surface waters in a small agricultural catchment. Environ Sci Technol 38(14):3827–3834

Lohse KA, Brooks PD, McIntosh JC, Meixner T, Huxman TE (2009) Interactions between biogeochemistry and hydrologic systems. Annu Rev Environ Resour 34:65–96

Mahmoud M, Tawfik A, Samhan F, El-Gohary F (2009) Sewage treatment using an integrated system consisting of anaerobic hybrid reactor (AHR) and downflow hanging sponge (DHS). Desalin Water Treat 4(1–3):168–176

Massoud MA, Tarhini A, Nasr JA (2009) Decentralized approaches to wastewater treatment and management: applicability in developing countries. J Environ Manag 90(1):652–659

Menezes CT, Heller L (2008) A method for prioritization of areas for pesticides surveillance on surface waters: a study in Minas Gerais, Brazil. Water Sci Technol 57(11):1693–1698

Mudd GM (2007) Global trends in gold mining: towards quantifying environmental and resource sustainability. Resour Policy 32(1):42–56

National Research Council (1993) Soil and water quality: an agenda for agriculture. National Academies Press

Okubo T, Onodera T, Uemura S, Yamaguchi T, Ohashi A, Harada H (2015) On-site evaluation of the performance of a full-scale down-flow hanging sponge reactor as a post-treatment process of an up-flow anaerobic sludge blanket reactor for treating sewage in India. Bioresour Technol 194:156–164

Salt DE, Smith RD, Raskin I (1998) Phytoremediation. Annu Rev Plant Biol 49(1):643–668

Sato N, Okubo T, Onodera T, Ohashi A, Harada H (2006) Prospects for a self-sustainable sewage treatment system: A case study on full-scale UASB system in India's Yamuna River Basin. J Environ Manag 80(3):198–207

Schulz M, Löffler D, Wagner M, Ternes TA (2008) Transformation of the X-ray contrast medium iopromide in soil and biological wastewater treatment. Environ Sci Technol 42(19):7207–7217
Schwarzenbach RP, Escher BI, Fenner K, Hofstetter TB, Johnson CA, Von Gunten U, Wehrli B (2006) The challenge of micropollutants in aquatic systems. Science 313(5790):1072–1077
Schwarzenbach RP, Egli T, Hofstetter TB, von Gunten U, Wehrli B (2010) Global water pollution and human health. Annu Rev Environ Resour 35:109–136
Schwarzman MR, Wilson MP (2009) New science for chemicals policy. Science 326(5956): 1065–1066
Smith AH, Lopipero PA, Bates MN, Steinmaus CM (2002) Arsenic epidemiology and drinking water standards. Science 296(5576):2145–2146
Tandukar M, Uemura S, Machdar I, Ohashi A, Harada H (2005) A low-cost municipal sewage treatment system with a combination of UASB and the "fourth-generation" downflow hanging sponge reactors. Water Sci Technol 52(1–2):323–329
Tandukar M, Uemura S, Ohashi A, Harada H (2006) Combining UASB and the "fourth generation" down-flow hanging sponge reactor for municipal wastewater treatment. Water Sci Technol 53(3):209–218
United Nations Environment Programme (2007) Global environment outlook GEO4: environment for development. http://www.eoearth.org/article/Global_Environment_Outlook_%28GEO4%29:Environment_for_Development_%28e-book%29
United States Environmental Protection Agency (2008) 2008 report on the environment. EPA/600/R-07/045F, National Center for Environmental Assessment, Washington, DC. http://www.epa.gov/roe/
Vorosmarty CJ, McIntyre PB, Gessner MO, Dudgeon D, Prusevich A, Green P, Davies PM (2010) Global threats to human water security and river biodiversity. Nature 467(7315):555–561
World Health Organization UNICEF (2006) Meeting the MDG drinking water and sanitation target: the urban and rural challenge of the decade. Geneva, Switz./New York

Chapter 3
Remediation of Contaminated Urban Streams: A Decentralized Ecological Wastewater Treatment Approach

Priyanka Jamwal

Abstract India is going through a period of rapid economic change leading to unplanned urbanization that has impacted the quality of the available water resources in cities and towns. The quality of water in most river stretches within cities and towns fall below category E often rendering river water unfit for any use. These rivers are mostly fed by open storm drains and streams that carry untreated and partially treated sewage and run off from urban and rural landscapes. According to the latest report by Central Pollution Control Board (CPCB), around 50% of the total sewage generated in India flows untreated into various rivers and water bodies. At most places, these water bodies are used as sources for irrigation posing a serious risk to the human health. Given the rapid pace of population growth in cities and the inability of centralized wastewater treatment infrastructure to keep pace with the exponentially increasing wastewater treatment demand, it is important to devise long-term sustainable water treatment and reclamation solutions that can be deployed at a catchment scale. In this context, decentralized in-stream water reclamation technologies present itself as an emerging low-cost, sustainable, and effective solution. These technologies employ ecological processes for the reclamation of a surface water body. In this chapter, we will present (a) a brief overview of the current state of Indian rivers in urban areas, (b) the role of pollution control boards in regulating the surface water quality, and (c) an approach for the deployment of decentralized in-stream treatment technologies for water reclamation.

P. Jamwal (✉)
Centre for Environment and Development, Ashoka Trust for Research in Ecology and the Environment (ATREE), Jakkur 560064, India
e-mail: priyanka.jamwal@atree.org

S. Bhattacharya et al. (eds.), *Water Remediation*, Energy, Environment, and Sustainability, https://doi.org/10.1007/978-981-10-7551-3_3

3.1 Introduction

An explosive urban growth throughout India continues to exert extreme pressure on urban water bodies impacting human health, natural ecosystems, and livelihoods. Use of contaminated surface water for irrigation is widespread especially in peri-urban areas that receive round the year flows of wastewater generated in urbanized areas. Many studies have shown that the long-term exposure of soil to contaminated wastewater causes loss of soil fertility and has a direct impact on the livelihoods of farmers (Mara et al. 2007). Studies have also shown that crops irrigated with contaminated wastewater tend to accumulate heavy metals thereby posing risk to human health (Chary et al. 2008). In addition, long-term exposure to contaminated water leads to the loss of aquatic biodiversity.

A recent report on the status of the water quality of Indian rivers stated that urban stretches of major Indian rivers are heavily polluted. These stretches are majorly located in the urban and peri-urban areas of Indian cities. The water quality of these rivers was tested for 29 parameters. Both the organic and inorganic contaminants such as biochemical oxygen demand (BOD) and heavy metals were above the recommended surface water quality standards prescribed for the propagation of aquatic life and fisheries (US EPA 2016).

The presence of heavy metals poses a major challenge in peri-urban landscapes where the livelihoods of farmers depend on the wastewater inflows from the cities. At present, the surface water used for irrigation is regulated for four conventional water quality parameters (pH, conductivity, boron, and sodium absorption ratio) (CPCB 2012). These parameters, however, are not comprehensive enough to represent the quality of irrigation water especially when surface water receives untreated/treated inflows from domestic and industrial sources (Jamwal et al. 2016).

Table 3.1 presents the impact poor surface water quality in urban and peri-urban areas. The long-term water quality monitoring and environmental impact assessment data have clearly established bioaccumulation of heavy metals in crops as well as humans. Heavy metals find their way to surface water bodies from either geogenic or anthropogenic sources. The contribution of geogenic sources is minimal; however, large quantities of heavy metal enter into water bodies through industrial and domestic wastewater discharges (anthropogenic sources). In spite of the continuous effort of the regulatory and pollution control boards to control/regulate wastewater inflows, the levels of heavy metals in surface water bodies' remain at alarmingly high levels.

3.2 Role of Government Agencies

In order to prevent, control and mitigate the pollution of surface water bodies, the National River Conservation Plan (NRCP) of India was launched in 1998. Figure 3.1 presents the framework adopted by NRCP to control and prevent surface

Table 3.1 State and impact of rivers stretches flowing through the urban habitation (Jamwal et al. 2016)

Rivers	Location (reference)	Nature of contamination	Impacts/implications
Musi river	Peri-urban area—Hyderabad city (Sekhar et al. 2005; Srinivasan and Reddy 2009)	Presence of high levels fecal coliform and heavy metals such as Zn, Cr, and Cu in water	Higher rates of morbidity exist in the irrigated area High levels of heavy metals found in blood and urine samples of people exposed to contaminated food, milk, and water in the irrigated areas
Hindon river	Ghaziabad city (Suthar et al. 2009, 2010)	High levels of heavy metals (Cd, Cu, Cr, Fe, Mn, Zn, and Pb) present in river sediments	Industrial sources are the major contributor of heavy metals to the river River is heavily polluted with cadmium and moderately polluted with manganese, zinc, and lead
Kasadri river	Taloja industrial estate, Maharashtra (Lokhande et al. 2011)	Heavy metal levels in river were much higher than the maximum permissible limits BOD was consistently greater than 30 mg/l	High level of chemical and biological pollution Effluents from industrial and domestic sources are the major contributor of pollution to river
Yamuna river	Delhi city (Sen et al. 2011; Jamwal et al. 2011) Agra city (Singh 2001; Sen et al. 2011)	High level of nickel cadmium and chromium in river Fecal coliform levels above permissible limits	Significant enrichment of these metals was observed in plants, fish, and river sediments In addition to domestic sewage, urban runoff also contributes to fecal coliform load in river
Ganga river	Kolkata city (Kar et al. 2008; Aktar et al. 2010)	Fe, Mn, Pb, and Ni exceeded the standard for drinking water	River water was not suitable for drinking purposes due to the excess concentrations of Fe, Mn, Pb, and Ni River water is not suitable for irrigation due to the excess concentration of Mn The excess heavy metal load in river water is attributed to the discharge of industrial effluents and municipal wastes, the geology of river bed and catchment area

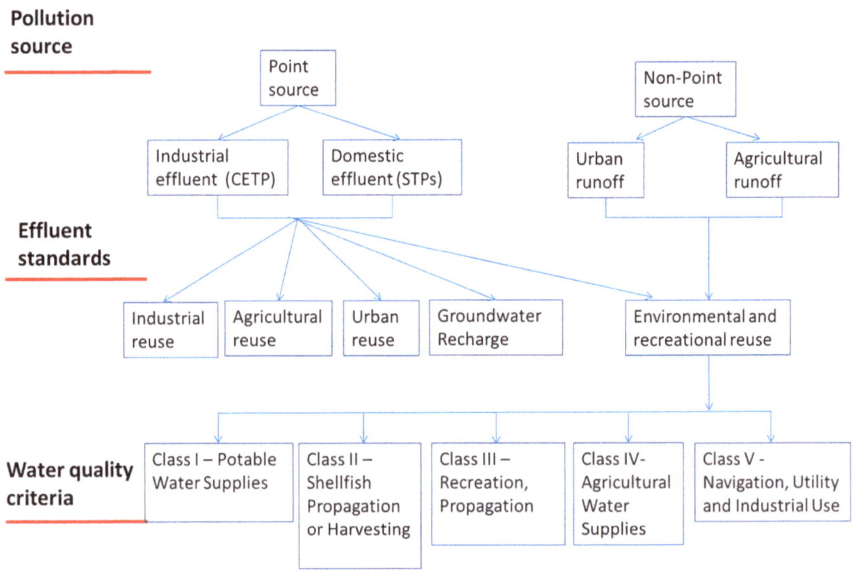

Fig. 3.1 National River Conservation framework to control pollution of the surface water bodies

water pollution. The first step of restoration/conservation plan is to identify the pollution sources. Contaminants within surface water can enter primarily from three types of sources (a) point sources, (b) nonpoint sources, and (c) distributed sources. As the name suggests, contaminants from point sources enter the water body at a particular location/point thereby making the process of controlling and regulating contaminants easy. In case of nonpoint sources, contaminants enter from multiple points, an example being urban runoff and agricultural runoff making the process of controlling and regulating contaminants difficult. The third categories of pollution source that are more relevant in the context of rapidly urbanizing cities/towns in developing countries are the distributed sources. Examples of such sources are small-scale industrial units, slums without sanitation facilities, car washing units, and other small-scale units that are distributed across the catchment. These sources discharge untreated dry and wet weather flows into the water bodies all year round.

Until now the focus of the state agencies has been to identify and control the pollution from the point sources. Funds are sanctioned by the state to control/reduce the contaminant inputs from domestic and industrial sources. For instance, in the year 2011, a total of 4919 million liters per day (MLD) of sewage treatment capacity was sanctioned in 20 states across India. A large number of domestic wastewater treatment plants were set up in all major cities and towns located on the banks of major Indian rivers. To mitigate industrial pollution, industries releasing BOD load more than 100 kg/day were mandated to install in-house effluent treatment units by law. For medium and small-scale industries, common effluent treatment plants (CETPs) were set up with the financial help of state pollution control boards (NRCP 2017).

Except for the Ganga Action Plan, none of the river action plans had focused attention/efforts toward mitigating pollution inputs from the nonpoint sources. Figure 3.2 shows the water quality trend of Ganga River at various locations in the state of West Bengal. The yearly trend shows no significant improvement in the water quality of river since the inception of the plan. Whereas the data from

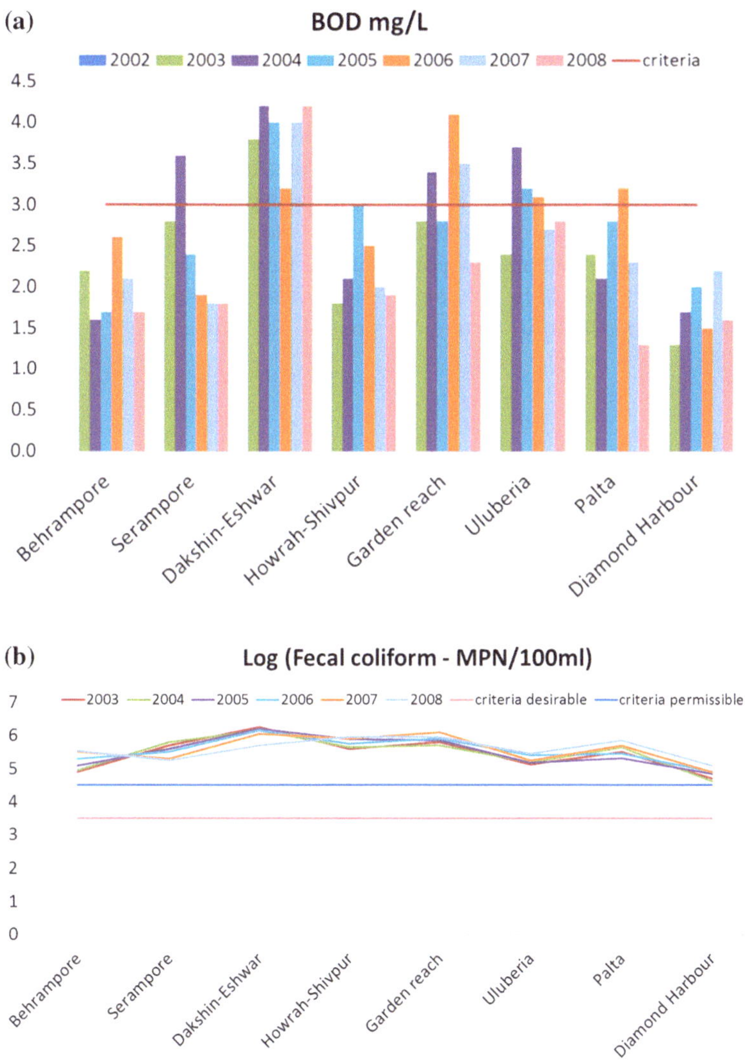

Fig. 3.2 Water quality status at various locations on Ganga River the state of West Bengal, **a** BOD and, **b** fecal coliform (CPCB 2015)

wastewater treatment plants (both Industrial and Domestic) show that the treated water meets the effluent discharge standards (CPCB 2013). The relevant question here is that if all STPs and CETPs meet effluent discharge standards then why rivers like the Ganga do not meet the desired surface water quality standards?

To find an answer to this question, we need an integrated approach to understand the factors that cause changes in the surface water quality. As stated above, anthropogenic activities in the catchment influence the quality of surface water bodies. Nations adopt different approaches to regulate pollution from various anthropogenic activities. For instance, Ireland adopted integrated catchment management approach to effectively manage pressures on their water sources at a catchment scale. The scale is large enough so that all the relevant information could be taken into account but small enough to ensure that people who live in the area can easily relate to their catchment. Experience around the world in addition to Ireland has shown that an integrated approach to managing individual catchments of an appropriate scale is necessary to protect and improve water resources. The integrated approach uses both the top-down (centralized treatment systems) and bottom-up solutions (best management practices) to control and regulate contaminant flux from both point and nonpoint sources. However, in developing countries like India, focus has been to employ top-down solution such as centralized wastewater treatment infrastructures to control contaminant flux into the water bodies.

3.3 Top-Down Approach

Larger infrastructural responses are being currently proposed and deployed by several city municipal authorities but are not scaling fast enough to keep pace with the daunting and shifting face of this challenge (Naveen Menezes 2016). In the year 2009, the total wastewater treatment capacity stood at 19,838 MLD which is approximately half of the wastewater generated in India. However, inadequate training and maintenance often limit the performance of existing large-scale, technologically intensive infrastructures such as sewage treatment plants (STP's), leading to only a part of the designed capacity being utilized (Jamwal et al. 2009, 2015) (Table 3.2). This highlights the need to focus on and invest in immediate, simple, and low-maintenance responses that can be implemented and tested bottom-up.

As discussed in the previous sections, in addition to point sources, nonpoint and distributed sources contribute significantly to the dry weather inflows into the surface water bodies. Figures 3.3 and 3.4 highlight the sources and the factors that contribute to the surface water quality. Figure 3.4 also lists down the effects of seasonality on the surface water quality. The surface water bodies in the cities and towns are usually fed by a network of small rivulets; open storm drains that carry urban runoff both during dry and wet weather. Treated/partially treated effluents from point sources and untreated flows from the distributed sources contribute

Table 3.2 Status wastewater generation and treatment capacity across Indian cities and towns (CPCB 2009)

	Metro cities (population > 1 million)	Class I cities (population 100,000–1 million)	Class II towns (population 50,000–100,000)
No. of cities	35	498	410
Sewage generated	15,644 MLD	35,558 MLD	2697 MLD
Sewage treatment capacity (installed)	8040 MLD	11,554 MLD	234 MLD
Treatment capacity (% of total sewage)	51%	32%	8%

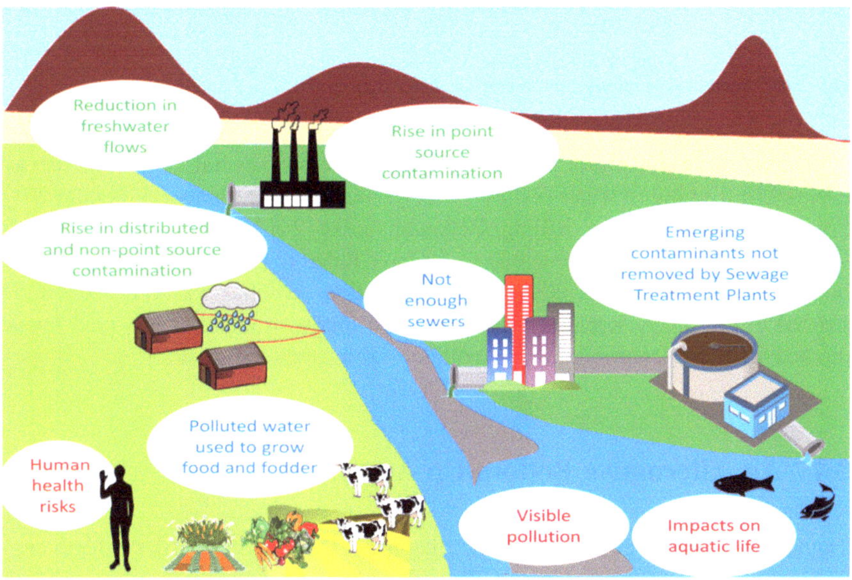

Fig. 3.3 Sources contributing to water quality of surface water bodies

significantly to the dry weather flows in streams that in turn are among the major reasons of surface water pollution.

There has been a collective civil society action across Indian cities to protect the quality of water bodies. The primary focus is to restore and protect the water bodies from encroachment. The actionable activities include removal of the debris around the lake, fencing of the lakes, protecting buffer zones, and ensuring that lake appears healthy.

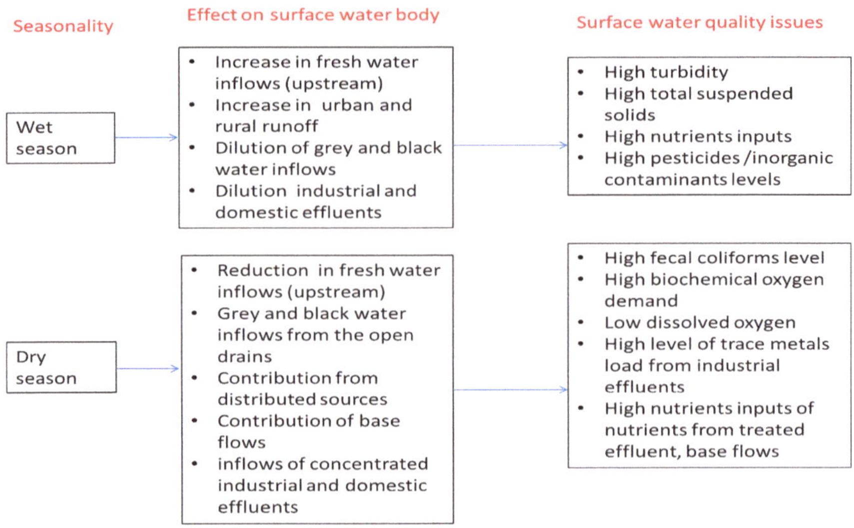

Fig. 3.4 Effect of seasonality on the surface water quality

In spite of all the efforts, cities have often witnessed several large-scale fish kill events and frothing in water bodies (Ulsoor Lake Fishkill: Plunder of Nature 2017; Pollution, Encroachment Killing Bellandur Lake 2015). This implies that the solution/interventions implemented so far have failed to protect the quality of water bodies. The reason is that the focus of restoration activities was very narrow (restricted to the lake/water body premises), and minimal attention was paid toward the activities that contribute to the pollution at a catchment scale.

3.4 An Alternative Bottom-Up Approach

There is sufficient evidence in the literature that illustrates the effectiveness in-stream treatment technologies (a bottom-up approach) for preventing and mitigating pollution from the distributed and nonpoint sources (Seidl et al. 2008; Biomatrix Water 2017). These in-stream solutions are cost-effective, robust low-maintenance systems that basically enhance the self-purification capacity of the stream and bring about a measurable positive impact on the quality of the receiving water body. These in-stream treatment systems are an innovative new way of tackling pollution by means of low-tech, affordable, and flexible interventions that fit within urban storm water channels and can target domestic effluents such as gray and black water, as well as partially treated industrial effluents, which are widespread and among the most common forms of wastewater contamination within the urban and peri-urban areas of India.

The following section presents the approach for the deployment of in-stream treatment systems at the catchment scale.

- *Hot Spot Identification*: Water quality monitoring should be carried out to identify the hot spots within the catchment, understand the relationship between land use, land cover, population density, social factors, and the pollution load in the stream. This would determine the number of in-stream (modular) systems required to be deployed at the catchment scale to achieve desired water quality.
- *Catchment Scale Deployment*: The in-stream treatment systems have a potential to produce an impact at the catchment scale. Many systems can be moored in sequence down a river at various points (e.g., Fig. 3.5), each treating the pollution load entering a particular section and feeding into the other downstream. Each in-stream system can be approximately 100 m (Modular) and can be deployed at the hot spots identified by water quality monitoring.
- *Leverage Local Materials*: The in-stream treatment technology should employ locally available filter materials, pallets of locally available plants, and microorganism to achieve in stream pollutant removal. The rubric of plants and microorganism should be deployed to enhance the effectiveness of treatment systems.
- *Models for Maximizing Impact*: After assessing the pollution loads and understanding the contribution of various factors on the contamination levels in stream, simulation/mathematical model should be developed to predict and determine the optimal locations and number of in-stream treatment technologies

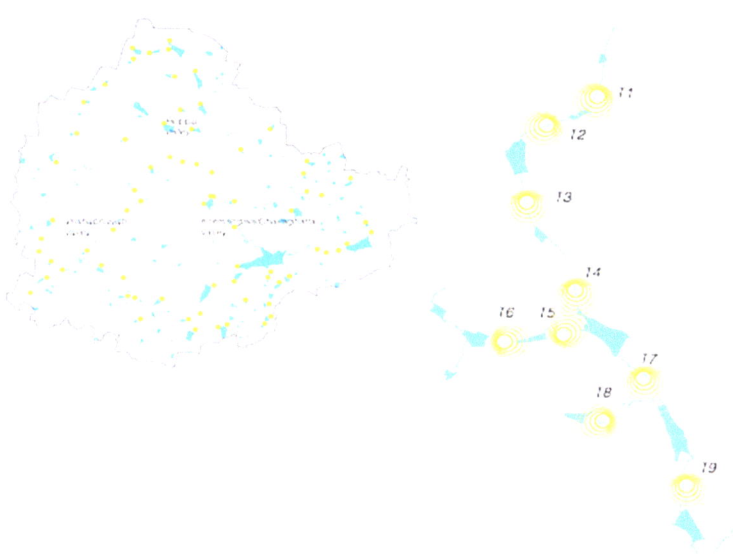

Fig. 3.5 Illustration of bottom-up approach at the catchment level. The rings show the optimal placement of in-stream treatment systems at the catchment scale

that are required to be deployed at the catchment scale to maximize the desired output (pollution reduction).

- *Policy Influence*: The model can then be presented to policy makers to emphasize the importance and effectiveness of decentralized and low-cost in-stream technologies in addressing the pollution of surface water bodies.

Table 3.3 summarizes the literature on the effectiveness of low-cost treatment technologies when deployed in the field and in the laboratory. Studies have shown that in addition to removing the biological parameters these systems have the potential to remove trace metals, pesticide, and nutrients from the wastewater.

Table 3.3 Effectiveness of various wetland/in-stream treatment technologies tested in laboratory or in field

Type of stream/ lab setup	Parameters	Technology	Efficiency
Lab setup (Vymazal 2009)	BOD, chemical oxygen demand (COD) COD nitrogen, phosphorus suspended solids (TSS)	Constructed wetlands with horizontal subsurface flow for various types of wastewater	BOD removal efficiency 80% for domestic wastewater and 60% for industrial wastewater. Significant reduction >70% was observed in nitrogen and phosphorus removal
The oil sands in northern Alberta, Canada. River and regional lakes (Allen 2008)	Dissolved solids, conductivity, pH, Na, Ca, Mg, Cl^-, bicarbonate, sulfate, ammonia, BOD and COD	Creating wet landscape environment	<10% efficiency in the reduction in the toxicity
Constructed wetlands Europe (Vohla et al. 2011)	Phosphorous	Tested different filter materials (natural materials, industrial byproducts, and man-made products) in constructed wetlands for removal of phosphorous	Highest P-removal capacities were reported for various industrial byproducts followed by natural materials and man-made filter media
Lab setup	Effectiveness of plants in the supply of oxygen, direct degradation of pollutants, role of microorganisms in pollutant degradation	Aquaculture systems, hydro botanical system, horizontal surface and subsurface flow systems, and vertical flow systems	Greater than 5% P-load is taken by plants. Nitrogen removal is between 5 and 10%. Carbon release 10–40% Different microorganisms as tested: for sulfur metabolism

(continued)

Table 3.3 (continued)

Type of stream/ lab setup	Parameters	Technology	Efficiency
Lab setup	BOD, total suspended solids (TSS), fecal coliform (FC), and total coliform (TC)	Horizontal subsurface flow wetlands—planted and unplanted	Planted wetlands were 10% more efficient than unplanted wetlands. BOD removal efficiency of unplanted wetland was 70%
Lab setup (Brix and Arias 2005)	TSS, BOD, COD, nitrogen, and Phosphorus	Vertical constructed wetland systems (planted filter beds)	Greater than 90% BOD and phosphorus removal were observed

3.5 Conclusion

The current assessment of the quality of water in major Indian rivers indicates that the urban stretches of these rivers are highly contaminated. Despite significant investments to control and prevent the pollution of water bodies, the pollution load in the rivers has shown an increasing trend over time. For the last few decades, the focus of NRCP has been to control the pollution from point sources. Little attention has been paid to prevent pollution inputs from various open storm drains that flow both during dry weather and wet weather seasons. Gray water, black water, and treated/untreated industrial effluents contribute to the dry weather flows. Whereas during wet weather, open drains receive additional inflows from paved surfaces. These untreated dry and wet weather flows are one of the major contributors to the surface water pollution. Deployment of decentralized in-stream treatment technologies at the catchment scale could be one of the potential solutions to address these issues. These systems do not offer the complete solution but could work together with the centralized wastewater treatment systems to achieve desired surface water quality all year round. These systems are not maintenance free and therefore would open doors for community participation and engagement that would eventually lead to sustainable model.

Acknowledgements The author is graciously thankful to her fellow researchers and Fulbright scholars at ATREE: Daniel Phillips and Kim Karlsrud from thecommonstudio.com for their contributions including research and numerous enriching discussions and pilot scale tests that have led to the development of this approach to deploy in-stream treatment technologies.

References

Aktar MdW, Paramasivam M, Ganguly M, Purkait S, Sengupta D (2010) Assessment and occurrence of various heavy metals in surface water of Ganga River around Kolkata: a study for toxicity and ecological impact. Environ Monit Assess 160(1–4):207–213

Allen EW (2008) Process water treatment in Canada's oil sands industry: I. Target pollutants and treatment objectives. J Environ Eng Sci 7(2):123–138

Biomatrix Water (2017) Water quality management. Biomatrix Water. http://www.biomatrixwater.com/water-quality-management/

Brix H, Arias CA (2005) The use of vertical flow constructed wetlands for on-site treatment of domestic wastewater: New Danish guidelines. Ecol Eng 25(5):491–500

Chary NS, Kamala CT, Samuel Suman Raj D (2008) Assessing risk of heavy metals from consuming food grown on sewage irrigated soils and food chain transfer. Ecotoxicol Environ Saf 69(3):513–524

CPCB (2009) Status of water supply, wastewater generation and treatment in class-I cities & class-II towns of India. CUPS/70/2009-10. Control of Urban Pollution Series. Central Pollution Control Board, Delhi. http://cpcb.nic.in/upload/NewItems/NewItem_153_Foreword.pdf

CPCB (2012) Water quality criteria. Web document. Central Pollution Control Board, New Delhi. http://cpcb.nic.in/Water_Quality_Criteria.php

CPCB (2013) Performance evaluation of sewage treatment plants under NRCD. Central Pollution Control Board, New Delhi

CPCB (2015) Polluted river stretches in India criteria and status. Accessed 4 Oct. http://cpcb.nic.in/upload/NewItems/NewItem_172_FinalPollutedStretches.pdf

Jamwal P, Mittal AK, Mouchel J-M (2009) Efficiency evaluation of sewage treatment plants with different technologies in Delhi (India). Environ Monit Assess 153(1–4):293–305

Jamwal P, Mittal AK, Mouchel J-M (2011) Point and non-point microbial source pollution: a case study of Delhi. Phys Chem Earth Parts A/B/C 36(12):490–499. https://doi.org/10.1016/j.pce.2008.09.005

Jamwal P, Md Zuhail T, Raje Urs P, Srinivasan V, Lele S (2015) Contribution of sewage treatment to pollution abatement of urban streams. Curr Sci 108(4):677–685

Jamwal P, Lele S, Menon M (2016) Rethinking Water Quality Standards in the Context of Urban Rivers. In Eighth INSEE Biennial Conference, Bengaluru Urbanization and the Environment, Indian society for ecological economics (INSEE), Bangalore

Kar D, Sur P, Mandai SK, Saha T, Kole RK (2008) Assessment of heavy metal pollution in surface water. Int J Environ Sci Technol 5(1):119–124

Lokhande RS, Singare PU, Pimple DS (2011) Quantification study of toxic heavy metals pollutants in sediment samples collected from Kasardi River flowing along the Taloja industrial area of Mumbai, India. N Y Sci J 4(9):66–71

Mara D, Sleigh P, Blumenthal U, Carr R (2007) Health risks in wastewater irrigation: comparing estimates from quantitative microbial risk analyses and epidemiological studies. J Water Health 5(1):39–50

Naveen Menezes (2016) BWSSB set to begin treatment for sick Bellandur Lake. In: Deccan Herald, 30 Apr. http://www.deccanherald.com/content/543530/bwssb8200set-begin-treatment-sick-bellandur.html

NRCP (2017) National River Conservation Directorate. http://www.moef.nic.in/sites/default/files/NRCD/More.html

Pollution, Encroachment Killing Bellandur Lake (2015) Deccan Herald. Accessed 26 May. http://www.deccanherald.com/content/479720/pollution-encroachment-killing-bellandur-lake.html

Seidl M, Von Sperling M, Castro P, Lima Dornelas F, Nascimento N (2008) Setup of an in stream treatment facility for urban creek revitalization, Belo Horizonte (Brazil). In: Proceedings of the 3rd SWITCH scientific meeting, BeloHorizonte, vol 30. http://switchurbanwater.lboro.ac.uk/outputs/pdfs/GEN_PAP_BH_Session7b_Instream_treatment_facility_for_creek_revitalisation.pdf

Sekhar C, Chary NS, Kamala CT, Shanker, Frank H (2005) Environmental pathway and risk assessment studies of the Musi River's heavy metal contamination—a case study. Hum Ecol Risk Assess 11 (6): 1217–1235

Sen I, Shandil A, Shrivastava VS (2011) Study for determination of heavy metals in fish species of the River Yamuna (Delhi) by inductively coupled plasma-optical emission spectroscopy (ICP-OES). Adv Appl Sci Res 2(2):161–166

Singh M (2001) Heavy metal pollution in freshly deposited sediments of the Yamuna River (the Ganges River Tributary): a case study from Delhi and Agra urban centres, India. Environ Geol 40(6):664–671

Srinivasan JT, Reddy VR (2009) Impact of irrigation water quality on human health: a case study in India. Ecol Econ 68(11):2800–2807

Suthar S, Nema AK, Chabukdhara M, Gupta SK (2009) Assessment of metals in water and sediments of Hindon River, India: impact of industrial and urban discharges. J Hazard Mater 171(1):1088–1095

Suthar S, Sharma J, Chabukdhara M, Nema AK (2010) Water quality assessment of River Hindon at Ghaziabad, India: impact of industrial and urban wastewater. Environ Monit Assess 165(1–4):103–112

Ulsoor Lake Fishkill: Plunder of Nature (2017) Deccan Herald. Accessed 31 Jan. http://www.deccanherald.com/content/534510/ulsoor-lake-fishkill-plunder-nature.html

US EPA, OW (2016) National recommended water quality criteria. Data and tools. Accessed 4 May. https://www.epa.gov/wqc/national-recommended-water-quality-criteria

Vohla C, Kõiv M, Bavor HJ, Chazarenc F, Mander Ü (2011) Filter materials for phosphorus removal from wastewater in treatment wetlands—a review. Ecol Eng 37(1):70–89

Vymazal J (2009) The use constructed wetlands with horizontal sub-surface flow for various types of wastewater. Ecol Eng 35(1):1–17

Part III
Use of Carbon Based Composites for Water Remediation

Chapter 4
Recent Advances in Carbon–Semiconductor Nanocomposites for Water Remediation

Kunal Mondal and Ankur Gupta

Abstract The demand of clean decontaminated water supply is increasing with the increase in the population of the world. As a result, the global need for more economic and efficient technologies for water remediation is also rising. The entry of nanotechnology into the wastewater treatment engineering exemplified a noteworthy advancement, and nanoparticles (NPs) have been exhaustively studied for possible remediation applications. However, the wastewater treatment with dispersed NP suspensions is still inadequate and to some extent antagonistic on the grounds of health and environmental safety, even as NP reaction mechanisms, ecotoxicity, and their transport properties are quiet under exploration. Theoretically, the development of porous nanocomposites containing nanoparticles to overcome these concerns offers the next logical step for developing functional nanomaterials and nanocomposites that are better investigated in the wastewater industry. This purpose of this review is to provide an overview of the range of carbon nanocomposites containing NP of semiconductor photocatalysts being technologically advanced, at the same time highlighting their limitations in practical uses. The review also briefly covers what further improvements are needed to optimize existing nanocomposite-mediated water remediation processes to successively accomplish viable maturity.

Keywords Carbon nanocomposites · Carbon/metal-oxide nanocomposites
Carbon/metal nanocomposites · Carbon/photocatalyst composites
Carbon nanofiber/metal-oxide photocatalyst · Wastewater remediation
UV and visible light photocatalysis · Water disinfection · Photocatalyst recycling

---

K. Mondal (✉)
Department of Chemical and Biomolecular Engineering, North Carolina State University, 911 Partners Way, Raleigh, NC 27695, USA
e-mail: kmondal@ncsu.edu

A. Gupta
School of Mechanical Sciences, Indian Institute of Technology Bhubaneswar, Bhubaneswar 752050, Odisha, India

S. Bhattacharya et al. (eds.), *Water Remediation*, Energy, Environment, and Sustainability, https://doi.org/10.1007/978-981-10-7551-3_4

4.1 Introduction

In recent times, an excessive research attention has been dedicated to photocatalytic elimination of organic pollutants and inactivation of microorganisms from wastewater using nanostructured photocatalysts (Fujishima et al. 2008; Tong et al. 2012; Linic et al. 2011; Jiang et al. 2017; Rtimi and Kiwi 2017; Bora and Mewada 2017; Park 2017; Wu et al. 2017a; Masih et al. 2017; Mohapatra and Parida 2017; Zhang et al. 2017; Cheng et al. 2017a; Wang et al. 2017a, b; Chen et al. 2011, 2017; Fang et al. 2017; Adamu et al. 2017; Nishikiori et al. 2017; Williams et al. 2008; Gaya and Abdullah 2008; Yoon et al. 2010; Ohno et al. 2004; Niu et al. 2012; Daneshvar et al. 2004). Among the photocatalysts, metal-oxide semiconductors, for example, titanium dioxide (TiO_2), zinc oxide (ZnO), zirconium dioxide (ZrO_2), tin oxide (SnO_2), tungsten trioxide (WO_3), silicon dioxide (SiO_2), have been considered most widely owing to their high photosensitivity, suitable bandgap (E_g = 2–4 eV) energies, and physicochemical stability (Park 2017; Asahi 2001; Zhao et al. 2004; Yan et al. 2009; Sakthivel et al. 2003; Wang et al. 1998; Yan et al. 2010; Kanakaraju et al. 2014; Chakrabarti and Dutta 2004; Permadani et al. 2016; Basahel et al. 2015; Prakash et al. 2016; Dominguez et al. 2017; Rey et al. 2014; Rashid et al. 2015). However, some problems can arise when those metal oxides are employed as a photocatalyst. For instance, photoinduced electron–hole pairs generated due to illumination can recombine rapidly, which causes lower photocatalytic efficacy (Miura et al. 2016; Nosaka et al. 1986; Sariciftci et al. 1994). On the other hand, since metal oxides have wide bandgap, few of them can only be energized with the exposure of ultraviolet (UV) radiation, unfortunately, which is less than 5% of total solar energy spectrum (Licht 2001; Ameta and Ameta 2016; Li and Wu 2015) Additionally, those nanostructured catalyst particles that tend to agglomerate during photocatalytic reaction in solution cause further decrease in overall photocatalytic efficiency. Then, it is imperative to develop photocatalytic materials that can be excited by light which is in visible spectrum. It has been reported that adding carbon nanophases to metal oxides can alter the properties of the semiconductor surfaces as well as their bandgap energies and shield the energy absorption band to the wavelength of visible light (Jafari et al. 2016; Lin et al. 2017; Khan et al. 2012). Carbon nanomorphologies such as carbon quantum dots, carbon spheres, carbon nanotubes, and graphene own unique electronic band structures and in some cases exceptional catalytic properties. For example, TiO_2 with carbon quantum dots turns as an electron trap, which helps faster electron–hole separation and eases interfacial electron transfer by lowering the chance of their recombination (Pan et al. 2015; Das et al. 2016; Cheng et al. 2017b).

Recently, carbon materials containing hierarchy in pore sizes have been established (Su et al. 2007; Katiyar et al. 2016; Mondal et al. 2016; Singh et al. 2017; Ren et al. 2013). Porous carbon nanofibers, carbon nanotubes, carbon nanospheres, and graphene oxides display higher adsorption ability and faster adsorption rate as compared to conventional porous carbon materials (such as activated carbon), due to the narrow and uniform pore structure (Mondal et al. 2017; Wei et al. 2016;

Liang et al. 2013; Neimark et al. 2003; Chen et al. 2016; Kumar et al. 2016). Moreover, shallow microspores exposed directly to the surface, which result in large adsorption capacity and ultrafast adsorption/desorption. Hierarchal pores where micro- and mesopores connected to macrospores makes easy transportation of reactant and products throughout the chemical reactions (Katiyar et al. 2016; Mondal et al. 2013).

The core focus of this chapter is to review the newly published works in the area of ultraviolet (UV) and visible light-induced photocatalysis of wastewater via photocatalyst/carbon nanocomposite. The summary of photocatalytic removal of contaminant dyes, wastewater-containing harmful bacteria, and volatile organic compounds (VOC) presented in environment demonstrated by various investigators using photocatalyst nanocomposite, and their shape and sizes, including information about the synthesis routes, are obtainable in Table 4.1. Entire consequences using different organic wastes exhibited that the use of photocatalysis directed to effective and whole degradation and formed less harmful products which are environmentally safe. However, partial degradation can have resulted in the creation of toxic by-products.

4.2 Carbon and Its Composite with Semiconductors for Water Remediation Application

4.2.1 Mechanism of Photocatalysis

When photocatalyst absorbs ultraviolet (UV) light from sunlight or artificial irradiated light source (artificial light source, e.g., fluorescent lamps), it yields pairs of electrons and holes. The electron in the valence band of semiconductor catalyst becomes excited when irradiated by radiation of light. The additional energy of this excited electron helps the electron to go to the conduction band of the catalyst leaving behind a positive hole. As a result, the negative electron (e^-) and positive hole (h^+) pair is created. This step is denoted as the semiconductor catalyst's 'photoexcitation' state. The difference in energy between the valence band and the conduction band is called the 'bandgap' energy.

However, the quantum efficiency of semiconductor photocatalysts is somewhat truncated because of the quick recombination of photocreated electron–hole pairs (Liu et al. 2012; Williams and Kamat 2009; Akhavan 2010; Dindar and Içli 2001). Also, the activity and stability of photocatalysts decreased intensely for the reason of photocorrosion throughout light illumination (Fan et al. 2012). Thus, to improve the photocatalytic efficacy, it is imperative to avoid the electron–hole from recombination. Newly, few works have been made on the heterojunction of nanostructured semiconductor photocatalyst synthesis with carbon nanomaterials such as carbon nanotubes, describing that the carbon nanotube/semiconductor composites can spread the usages and reinforce the virgin characteristics of the

Table 4.1 Carbon–photocatalyst composite and their environmental remediation applications

Composites	Shape and size of nanophocatalyst (nm)	Fabrication method	Light source	Application	Reference
Carbon–ZnO	Spherical ZnO nanoparticles distributed into carbon, size 8–12	Hydrothermal process	Visible	Photocatalysis of Rhodamine B dyes	Akir et al. (2017)
Carbon–TiO_2	10–40 nm-sized TiO_2 particles into amorphous carbon matrix	One-step carbonization method	UV–Visible	Photocatalysis of Rhodamine B dyes	Teng e al. (2014)
Carbon–SiO_2	Carbon-doped mesoporous 20–30 nm SiO_2 particles	Thermal decomposition of tetraethylorthosilicate	Near-UV light illumination	Photocatalysis of Rhodamine B dyes	Zhang et al. (2013b)
Carbon–Sn–SnO_2	Micron–submicron-sized rod-sphere-shaped Sn–SnO_2 distributed into carbon	Epoxide-assisted sol–gel method	UV	Photochemical degradation of Rhodamine B	Hung-Low et al. (2016)
Carbon–ZrO_2	4.8–39.3 nm ZrO_2 smooth particles distributed in the carbon xerogel framework	Sol–gel synthesis	Visible	Photocatalytic degradation of Orange G	Bailón-García et al. (2017)
Graphene–ZnO–Au	Graphene–zinc oxide nanospheres (G–ZnO NSs; average diameter of 45.3 ± 3.7 nm)	Hydrothermal process	UV	Photocatalytic reduction of nitrobenzene	Roy et al. (2013)
Graphene–TiO_2–Au	Au nanoparticles with round shape and about 10 nm in size were loaded on TiO_2 particles uniformly	Hydrothermal process	Visible	Photocatalytic reduction of 2,4-dichlorophenol(2,4-DCP)	Dang et al. (2014)
Graphene–SiO_2/Cu_2O	5–10 nm Cu_2O nanoparticles, 150–200 nm TiO_2 rods in mesoporous graphene	Self-assembly method with tetraethyl orthosilicate (TEOS)	Visible	Photocatalytic reduction of Rhodamine B, methylene blue trihydrate, and reactive black B	Nguyen et al. (2017)

(continued)

Table 4.1 (continued)

Composites	Shape and size of nanophocatalyst (nm)	Fabrication method	Light source	Application	Reference
Reduced graphene oxide–ZnO–TiO_2	Curled and corrugated RGO nanosheets and ZnO, TiO_2 displayed the sheet nanostructure and particle like, respectively	Microwave-assisted reduction of graphene oxide in ZnO precursor solution with TiO_2 suspension	UV	Reduction of Cr(VI)	Liu et al. (2013)
Graphdiyne–ZnO	100–200 nm ZnO nanoparticles bounded on 2D graphdiyne nanosheets	Hydrothermal process	UV	Photodegradation of two azo dyes, viz. (i) methylene blue and (ii) Rhodamine B	Thangavel et al. (2015)
Carbon nanofibers/TiO_2	Smooth fiber about several microns in length and 500–600 nm in diameter with TiO_2 nanorods about 100–200 nm in length and 5–10 nm in diameter	Electrospinning technique followed by hydrothermal treatment	Visible	Photodegradation of congo red (CR), methylene blue (MB), methyl orange (MO), and eosin red (ER)	Li et al. (2015a)
Carbon film–TiO_2	10-μm thin film with monodispersed carbon-doped TiO_2 particulates with a size of around 5 nm	Sol–gel process combined with hydrothermal treatment	UV and visible	Photo degradation of reactive brilliant red X-3B	Lin et al. (2011)
Fluorinated polymeric carbon nitride	Slate-like, stacked lamellar texture and broken up, and a more open texture observed due to fluorination	Fluorination route of polymeric carbon nitrides	Visible	Photocatalytic activity in both hydrogen production from water and oxidization of benzene to phenol	Wang et al. (2010)
Carbon nanoparticles–gold and platinum nanoparticle	Sub-10 nm gold and platinum nanoparticle-functionalized carbon particles	Surface functionalization in aqueous medium	Visible	Formation of formic acid from CO_2 photoconversion	Cao et al. (2011)

(continued)

Table 4.1 (continued)

Composites	Shape and size of nanophocatalyst (nm)	Fabrication method	Light source	Application	Reference
TiO_2–graphene –carbon nanotube –fullerene (C60) nanocomposite	Particle shape and the size of the TiO_2 nanoparticles in TiO_2–carbon composites are all about 10 nm	Sol–gel process along with hydrothermal post-treatment	Visible	Photocatalytic oxidation of benzyl alcohol	Yang et al. (2013)
3D carbon felt–TiO_2 monoliths	Carbon fibers with diameters ranging from 10 to 15 μm with TiO_2 nanoparticles sizes of several nanometers	Solvothermal process	UV	Photocatalytic performance on methyl orange degradation	Shen et al. (2016)
Carbon nanotubes–TiO_2 nanotubes	50-nm carbon nanotubes with 10-nm TiO_2 nanotubes	Mixing carbon nanotubes/TiO_2 nanotubes nitric acid solution	UV	Photodegradation of methyl orange	Jiang et al. (2013)
Amorphous carbon–TiO_2	10–20-nm TiO_2 particles in amorphous carbon	Solvothermal method	Visible	Photochemical degradation of Rhodamine B	Sha et al. (2016)
Graphite–graphene oxide	Graphite and graphene oxide in the form of sheet	Solvothermal	UV	4-chlorophenol removal from water	Bustos-Ramírez et al. (2015)
Carbon–indium oxide (In_2O_3) nanobelts	Lengths of 1–2 μm and widths of 50–100 nm (In_2O_3) nanobelts coated by a 5-nm-thick carbon	Hydrothermal process	Solar irradiation	Photocatalytic reduction of CO_2 to CO and CH_4 with water	Pan et al. (2017)
N-doped carbon quantum dots–TiO_2	Quasi-spherical carbon quantum dots diameters ranging from 6 to 13 nm and carbon quantum dots/TiO_2 diameters around 20–30 nm	Microwave-assisted method and then hydrothermally combined	UV and visible	Photocatalytic removal of NO_x under UV and visible light and methylene blue under UV light	Martins et al. (2016)

(continued)

Table 4.1 (continued)

Composites	Shape and size of nanophocatalyst (nm)	Fabrication method	Light source	Application	Reference
Activated carbon–TiO_2	Anatase TiO_2 particles with a crystal size of 10–20 nm deposited homogeneously on activated carbon surface	Sol–gel method	UV	Degradation of Rhodamine B	Xing et al. (2016)
Carbon aerogel–TiO_2	TiO_2 film composed of granular particles of size was about 20 nm	Ambient drying and sol–gel method	UV	Photocatalysis-enhanced electrosorption treatment of high-concentration alizarin red wastewater	Jin et al. (2011)
Vertically aligned multiwalled carbon nanotube array–TiO_2	TiO_2 nanoparticles with 100–200 nm size covered surface of multiwalled carbon nanotubes of 5–8 μm length and 60 nm diameter	Spray pyrolysis and coating	UV and visible	Photocatalytic removal of enterohemorrhagic pathogen *Escherichia coli* O157:H7	Oza et al. (2013)
Carbon–Ag–TiO_2	Particle size around 10–20 nm	Sol–gel method	Visible	Photocatalytic removal of methyl orange	Nyamukamba et al. (2017)
Carbon–nitrogen-doped TiO_2–SiO_2	C–TiO_2–SiO_2 composed with spherical particles with size in 14.7 nm range and TiO_2 with 18.9 nm	Sol–gel method	UV and visible	Photooxidation of methyl orange and phenol red	Mungondori and Tichagw (2012)
Carbon–nanotube–Ag_3PO_4 in Pickering emulsions	Ag_3PO_4 nanoparticles of 312.8 ± 94.6 nm size	Emulsion based on solution-dispersion and surface-immobilization	Visible	Photooxidation of methylene blue and oxygen evolution	Zhai et al. (2013)

(continued)

Table 4.1 (continued)

Composites	Shape and size of nanophocatalyst (nm)	Fabrication method	Light source	Application	Reference
Carbon nanofibers–Ag–TiO_2	Silver nanoparticles with uniform diameters in the range of 6–18 nm dispersed on Ag–TiO_2/CNF composites of several 100 nm diameters	Electrospinning followed by carbonization	Visible	Photooxidation of methylene blue	Kim et al. (2011)
Multiwall carbon nanotubes–TiO_2–SiO_2	Multiwall carbon nanotubes with outer diameter 20–30 nm, length of 10–20 mm and anatase TiO_2 and SiO_2 of 7–8 nm in size	Sol–gel method and co-precipitation method	UVA	Photocatalytic removal of bisphenol A and carbamazepine from water solution	Czech and Buda (2015)
Multiwalled carbon nanotubes–metal-doped ZnO nanohybrid	Multiwalled carbon nanotubes of diameters 20–50 nm uniformly modified by nanoparticles with sizes of the range 10–20 nm	Co-precipitation method	UV	Photocatalytic removal of methyl orange	Chen et al. (2013)
Graphene nanoplatelets–SiO_2	Spherical monodispersed SiO_2 particles with the diameter ranging from 230 to 260 nm decorated on few layers of graphene sheets	Solvothermal chemical route	UV	Photodegradation of methyl orange and bacterial inhibition of the growth of Staphylococcus aureus, *Escherichia coli* and *Pseudomonas aeruginosa*	Arshad et al. (2017)

distinct semiconductors and CNTs since carbon nanostructures grabbed the responsibility role of electron foraging (Fan et al. 2012; Zhang et al. 2009; Woan et al. 2009). In this context, Ali et al. (2016a) demonstrated the fabrication of a mesoporous few-layer reduced graphene oxide and nickel oxide (rGO–NiO) nanocomposites, and their SEM and TEM images are shown in Figs. 4.1 and 4.2, respectively. They have demonstrated that NiO nanoparticles were distributed on rGO surface or between the rGO layers causing to new type of nanocomposites with improved catalytic, electrochemical, and mechanical properties.

Nevertheless, some glitches still hamper the additional raise of efficiency of the existing CNTs/semiconductor composites, for example, the noteworthy reduction in the absorptivity through photocatalysis, the reduction in the intensity of light till received at the surface of the catalyst, and the problem of reproducibility owing to the fabrication and treatment discrepancy (Woan et al. 2009; Matsuda et al. 2013). In contrast with carbon nanotube, graphene is unquestionably a substitute because

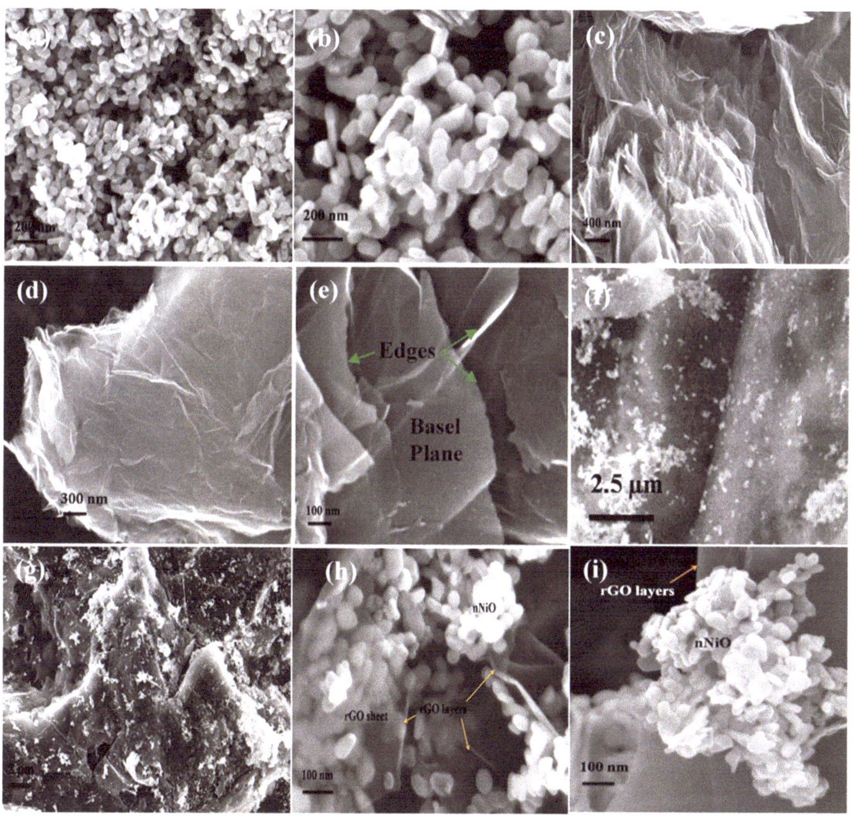

Fig. 4.1 **a**, **b** FESEM micrographs of NiO nanorods, **c**–**e** rGO sheets, **f**, **g** rGO–NiO composite, and **h**, **i** high-resolution view of rGO–NiO composite. Reprinted with permission from Ali et al. (2016a)

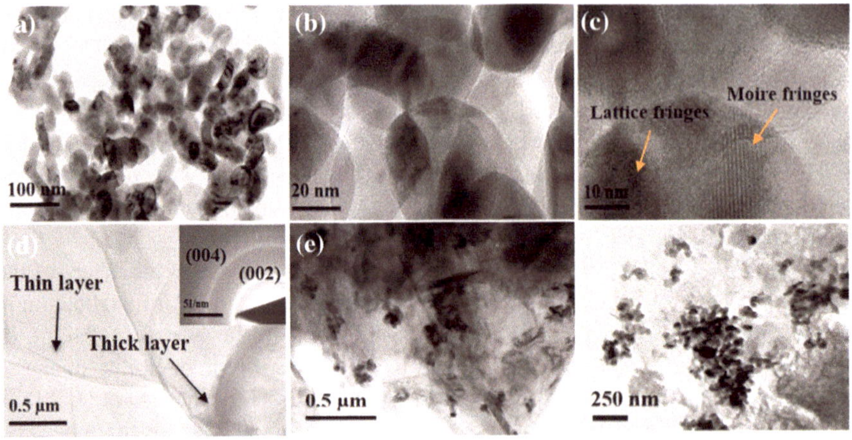

Fig. 4.2 **a**, **b** HRTEM micrographs of NiO nanorods, **c** atomic scale image of NiO nanorods, **d** rGO sheets, **e** rGO–NiO composite, and **f** higher resolution view of rGO–NiO composite. The inset of (**d**) shows selected area electron diffraction pattern (SAED) of rGO. Reprinted with permission from Ali et al. (2016a)

of its countless advantages, for instance, high transparency, zero bandgap energy, and highly electronic conductivity for storage and transportation of electrons (Allen et al. 2010; Rao et al. 2009) by reason of its one-atom-thick large theoretical specific surface area (McAllister et al. 2007). Likewise, the property of surface of graphene can be easily regulated by introducing simple chemical treatment, which permits its application in composites. Therefore, the marriage of semiconductor photocatalysts and carbon nanostructures is capable of showing outstanding transparency, absorptivity, conductivity, and formability, which could simplify the effective photocatalytic degradation of contaminants in wastewater. To understand the mechanism of photocatalysis enhancements and to design new carbon–semiconductor-based heterojunction composites for ultraviolet and visible light photocatalysis, it is essential to explore the interaction of electron–hole pairs involved in such carbon–semiconductor heterojunctions. Figure 4.3 shows the wastewater treatment using carbon–photocatalyst nanocomposite.

The photocatalytic decomposition of organic pollutant present in wastewater with the presence of UV light can be defined as follows (Mondal et al. 2013, 2014; Mondal 2017; Mondal and Moitra 2017; Singh et al. 2013, 2016; Gupta et al. 2015; Schneider et al. 2014):

$$\text{Carbon-photocatalyst composite} + \text{light energy } (hv) \rightarrow \text{electron } (e^-) + \text{hole } (h^+) \quad (4.1)$$

$$h^+ + H_2O \rightarrow H^+ + OH^{\cdot} \quad (4.2)$$

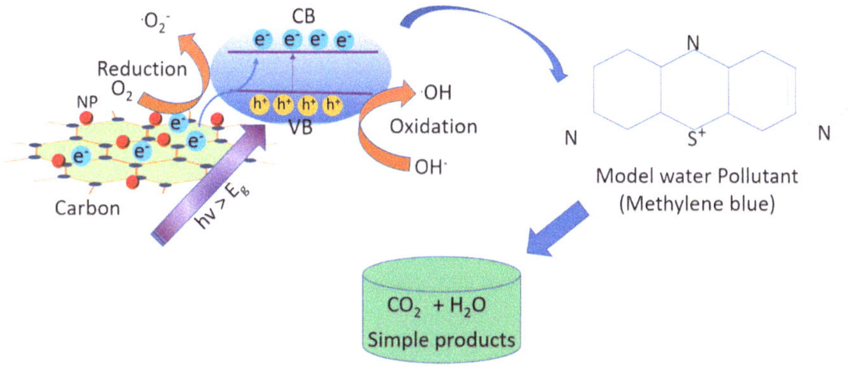

Fig. 4.3 Wastewater treatment using carbon–photocatalyst nanocomposite

$$h^+ + OH^- \rightarrow OH^{\cdot} \tag{4.3}$$

$$e^- + O_2 \rightarrow O_2^- \tag{4.4}$$

$$2e^- + O_2 + 2H^+ \rightarrow H_2O_2 \tag{4.5}$$

$$e^- + H_2O_2 \rightarrow OH^{\cdot} + OH^- \tag{4.6}$$

$$\text{Wastewater with pollutant} + OH^{\cdot} \rightarrow \text{clean water and degradation products} \tag{4.7}$$

$$\text{Wastewater with pollutant} + \text{Carbon-photocatalyst composite } (h^+) \rightarrow \text{oxidation products} \tag{4.8}$$

$$\text{Wastewater with pollutant} + \text{Carbon-photocatalyst composite } (e^-) \rightarrow \text{reduction products} \tag{4.9}$$

The graphene cannot interact with photocatalyst via chemical bonding. This restricts the transfer of charge particles between the graphene and photocatalyst and results in the poor photocatalysis. Yang and his coworkers have shown that the integration is possible between graphene and photocatalyst via the functional groups present on the surface of the graphene sheets when they become a photocatalyst/graphene nanocomposite (Fan et al. 2012). The existence of carbon nanostructures in the composite does not hamper the crystal structure of photocatalyst, and such interface can be accountable for the improved photocatalysis of photocatalyst/graphene. Moreover, the electronic band structure of semiconductor photocatalyst and carbon can help in such a way that they can easily undertake process of charge transfer under UV illumination with acceptor or donor electron present in the composite.

4.2.2 Photocatalytic Water Remediation

Recently, Xu et al. (2016) have fabricated ZnO–graphene composites from a simple chemical corrosion route and tested for their photocatalytic act on the decomposition of methyl orange dye and water splitting. Their finding displayed that the ZnO–graphene composites have a porous network assembly, which is an advantage to the adsorption and transfer of the mass of dye and oxygen molecules. They observed a decomposition rate of 87% in under visible light irradiation in 3 h with 8 mg ZnO–graphene photocatalyst and the water splitting rate up to 4.35 mmol g^{-1} h^{-1}. The better photocatalytic efficiency was credited to the porous framework, narrow band structure, and excellent charge transfer properties of the nanocomposites. The occurrence of graphene promoted the fast photocreated electron–hole separation and suppressed the recombination of charge carrier, which results in an enhanced photocatalysis.

A macroporous thin film of titanium dioxide/carbon hybrid was prepared directly through spin coating technique by Mondal et al. (2013). The TiO_2/carbon hybrid porous films were used to decompose a model pollutant dye Rhodamine B in aqueous medium, and the results exhibit an excellent photocatalytic activity. The composite film demonstrates its photocatalytic activity due to the existence of TiO_2 nanopartciles, while the porous carbon support offers an imperative role in bypassing the aggregation of the photocatalytic particles throughout the photocatalysis reaction and its microporous network helps in unaffected mass transport.

In another work, Lee et al. (2012) have reported an efficient and environmentally friendly biomimetic mineralization of titanium dioxide at the graphitic carbon surface, which created a TiO_2/carbon hybrid structure without any complex surface treatment or interfacial adhesive layer. They also reported that the direct contact of the surface of CNT and TiO_2 nanoshell presented a new carbon energy state in the TiO_2 band structure and in that way effectively lowered the bandgap energy of the photocatalyst and thereby demonstrates efficient visible light photocatalysis. Few years ago, Sharma's group (Mondal et al. 2014) has developed mesoporous structure of carbon-doped anatase TiO_2 nanofiber via electrospinning technique as shown in Fig. 4.4. They have demonstrated an effective and reusable photocatalytic activity of partially aligned pure electrospun nanofibers of TiO_2 and residual carbon-containing TiO_2 fibers with an average fiber diameter of $\sim$40 nm in the photocatalytic removal of polycyclic aromatic hydrocarbon dye, naphthalene. The degradation of the pollutant solution was analyzed through a UV-Vis spectrophotometer by measuring its absorbance peak at 275 nm. UV–Visible absorption spectra of naphthalene solution at various time intervals with a loaded composite fiber are shown in Fig. 4.5. They have also reported that small carbon residue $\sim$2.54% comprising TiO_2 nanofiber was found two times more efficient as the virgin TiO_2 nanofiber in photodegradation; however, the efficiency weakened at higher carbon loading into the semiconductor.

The interfacing carbon materials can increase the ability of the visible light absorption of semiconductor photocatalyst. In this context, Li et al. (2015b) have

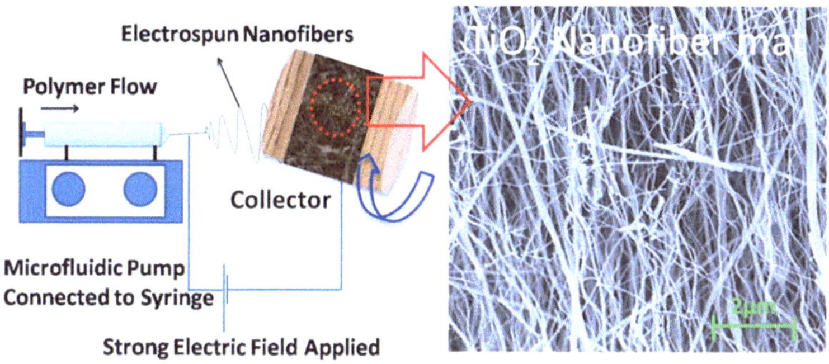

Fig. 4.4 Electrospinning setup for the synthesis of aligned carbon-doped TiO_2 nanofibers. Reprinted with permission from Mondal et al. (2014)

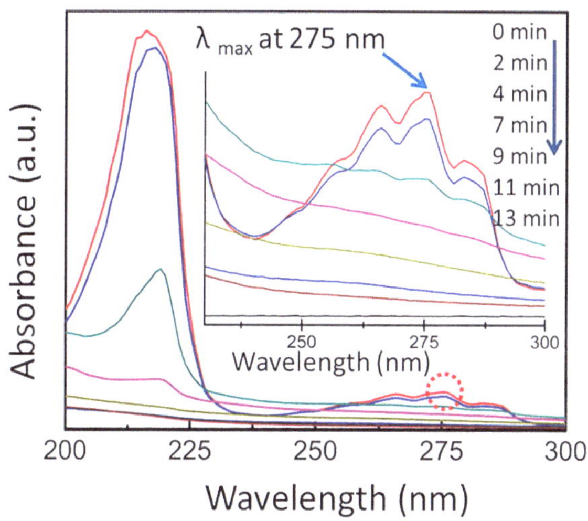

Fig. 4.5 UV–Vis absorption spectra for naphthalene degradation by carbon-doped TiO_2 nanofibers at different time intervals. The inset displays magnified λ_{max} at 275 nm wavelength. Reprinted with permission from Mondal et al. (2014)

synthesized carbon spheres loaded Cu_2O/Cu composites as shown in Fig. 4.6. They have demonstrated that the composite photocatalyst material can be used to degrade dye wastewater or other organic pollutant. Recently, Wu et al. (2017b) synthesized core–shell structure of TiO_2–carbon sphere where TiO_2 is anchored on carbon spheres via hydrothermal polymerization of sucrose to bathochromically enlarge its light-response area. They have shown that owing to the interface formation between the two materials, the composite accomplished an improved visible light absorption property compared to pristine TiO_2. The composite efficiently degraded organic pollutants, for example, methyl orange, bisphenol A, and Oseltamivir, under visible light irradiation.

Likewise, a series of applications for environmental remediation using carbon/photocatalyst nanostructures demonstrate the special advantages of their composite

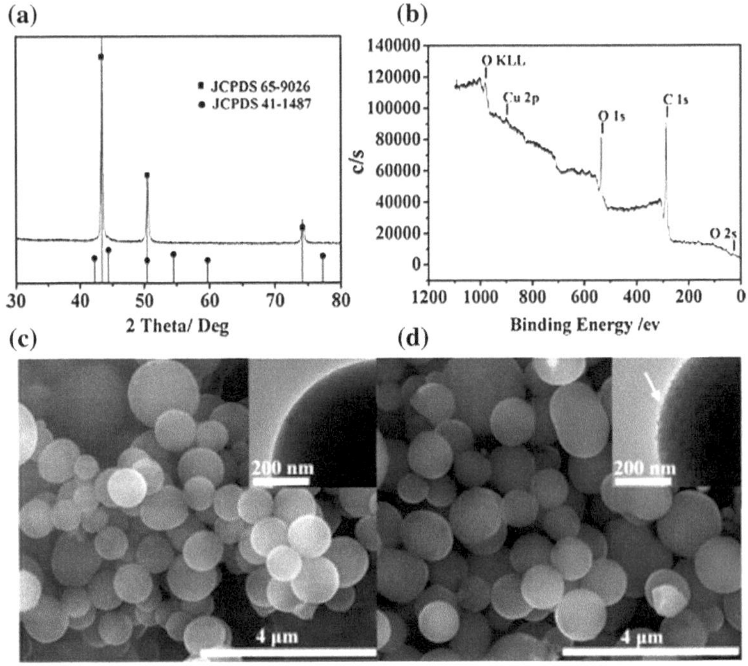

Fig. 4.6 **a** XRD analysis, **b** XPS analysis of Cu_2O-/Cu-loaded carbon spheres, **c** SEM, and **d** HRTEM micrographs. The inset shows individual of carbon sphere and Cu_2O-/Cu-loaded carbon sphere. Reprinted from Li et al. (2015)

morphologies and easily accessible high surface areas (Peraltahernandez et al. 2007; Yang et al. 2006; Sampaio et al. 2011; Chang et al. 2015; Fernando et al. 2015; Yi et al. 2016; Velasco et al. 2013; Xu et al. 2015; Bakos et al. 2016; Yang et al. 2014; Gu et al. 2017; Shi et al. 2008; Dey et al. 2011; Jo et al. 2014; Yao et al. 2014; Saleh 2013; Herring et al. 2012; Cai et al. 2016; Hosseini and Babaei 2016; Réti et al. 2017; Li et al. 2015c; Krishnamoorthy et al. 2011; Ji et al. 2013; Omar et al. 2014).

4.3 Water Disinfection Application

As a consequence of the extensive usage of antibiotics and the rise of extra resistant and poisonous strains of microorganisms, there is an instant requirement to discover alternate sterilization methods and technologies. In this context, decontamination of bacteria-containing wastewater vis photocatalysis is a conceptually simple and promising technology and has achieved considerable research consideration in recent years (Gumy et al. 2006; Kalt et al. 2014; Pelaez et al. 2012; Dong et al. 2015). The earlier work on disinfection and bacterial cell killing using photocatalytic process has been reviewed (Blake et al. 1999). Various photocatalysts have been used for the sterilization of microbial cells, together with organic pollutant removal from wastewater. There was one of the earlier reports of decontamination of wastewater using TiO_2/Pt nanoparticles for a variety of microorganisms, such as Lactobacillus acidophilus (bacteria), Saccharomyces cerevisiae (yeast), and *Escherichia coli* (bacteria) by Matsunaga et al. (1985).

They also fruitfully constructed a real photocatalytic device based on immobilized titanium dioxide powder on membrane made from acetyl cellulose. A suspension containing *E. coli bacteria* passing through this membrane was entirely disinfected (Matsunaga et al. 1988). The discoveries of Matsunaga et al. (1988) formed an innovative path for disinfection. This was caused in efforts to include this innovative photocatalytic technique for sterilizing drinking water and eliminating odor and bioaerosols from air (Byrne et al. 1998; Goswami et al. 1997; Salih 2002; Ireland et al. 1993; Saito et al. 1992; Stevenson et al. 1997; Wei et al. 1994). Nearly two decades ago, Yu et al. (2005) have shown decontamination of the polluted water comprising *Micrococcus lylae* (Gram-positive bacterium) using sulfur-doped nanocrystalline TiO_2.

The bactericidal activity of semiconductor-mediated photocatalysis has been reported by many authors (Dunlop et al. 2002; Maness et al. 1999; Sunada et al. 1998). Although there is a plenty of information available demonstrating the effective use of photocatalyst particles for removing biocontainments, the basic underlying mechanism for the photocatalytic disinfection is not studied well. Therefore, a detailed knowledge on the disinfection is necessary and indispensable to create a tactic and implement the expertise in a real scheme to proficiently execute an extensive range of bacteria. It is supposed that direct photoinduced chemical oxidation of dimeric form of intracellular coenzyme-A is a main reason of reductions in respiratory systems that directed to the death of bacterial (Matsunaga et al. 1985, 1988). Saito et al. (1992) anticipated that the photochemical reaction by the photocatalyst triggered destruction of the membrane and wall of the cell that leads to cell death. Conceivably, more direct indication was observed on the damage of the outer membrane and was reported by Sunada et al. (1998). They have studied *E. coli* and showed that the endotoxin, which is a vital constituent of the outer cell membrane, was demolished under photocatalytic environments.

Figure 4.7 describes the disinfection mechanism using carbon–photocatalyst composite. Hydroxyl radicals produced by the photocatalyst upon exposure of light energy are very strong oxidants (Mills and Le Hunte 1997). These radicals are also very short survived because of their high levels of reactivity. When light-illuminated photocatalyst nanostructures come in direct contact with or proximity to the microorganisms, the bacterial surface is the key target of the primary oxidative dose. The vulnerability of polyunsaturated phospholipids, which are an essential module of the bacterial cell membrane, to attack by reactive oxygen species (ROS) has been well understood (Gutteridge 1995; Sies 1985). Therefore, the damage of membrane assembly and membrane functions is the main basis of cell damage when catalysts occupied the outsides of microorganism cells.

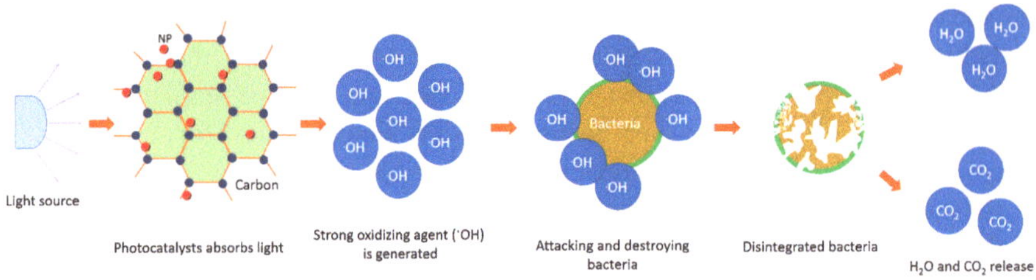

Fig. 4.7 Photocatalytic water disinfection mechanism

Recently, Sunada et al. (1998) have investigated the special features of the antibacterial effect for a transparent TiO_2 thin film for the photocatalytic degradation of endotoxin, which is a pyrogenic constituent of *E. coli*, as well as its bactericidal activity. Fe/ZnO nanoparticle-assisted solar-photocatalytic disinfection of multidrug-resistant *E. coli* isolated from wastewater of a rural healthcare center in synthetic as well as natural water systems was reported by Das et al. (2017). An efficient photocatalytic decontamination of *E. coli* O157:H7 bacteria was accomplished by Ouyang et al. (2016) using a C70-modified titanium dioxide (C70-TiO_2) hybrid photocatalyst below a visible light radiation source. Their disinfection experimentations exhibited that 73% of bacteria died in 2 h with a high decontamination rate (rate constant $k = 0.01\ \text{min}^{-1}$), which is three times more than that measured for TiO_2 alone. Enhanced visible light-induced photocatalytic disinfection of bacteria cells by carbon-sensitized N_2-doped TiO_2 composite was also studied (Li et al. 2007). Recently, it has been well demonstrated by Binas and coworkers that the elimination of nitrogen oxide gases in volatile organic compounds at typical urban/indoor air and disinfection against various microorganisms, including the most persistent in the environment such as viruses, bacterial spores, and protozoan cysts by the use of TiO_2-mediated photocatalysis is achievable (Binas et al. 2017). The research on a new type of heterogeneous photocatalysis system for decontamination of bacteria from polluted water using tetrakis (4-carboxyphenyl)porphyrin-sensitized titania–graphene nanocomposite has been reported by Rahimi et al. (2015). It was also shown that the sensitization of the titania–graphene nanocomposite with porphyrin is proficient for killing microorganisms with reasonable high yields in visible light radiation.

4.4 Other Applications

Carbon nanomaterials along with semiconductor photocatalysts are also included in numerous applications: textiles (Lee et al. 2014), medical (Ji et al. 2012), bioimplants (Dwivedi et al. 2011), biosensing (Hu et al. 2011), drug delivery (Sawant et al. 2016), tissue engineering (Salarian et al. 2014), energy generation (Shi et al. 2012), and storage (Luo et al. 2012), etc. (Miyazaki et al. 2008; Yan and McCreery

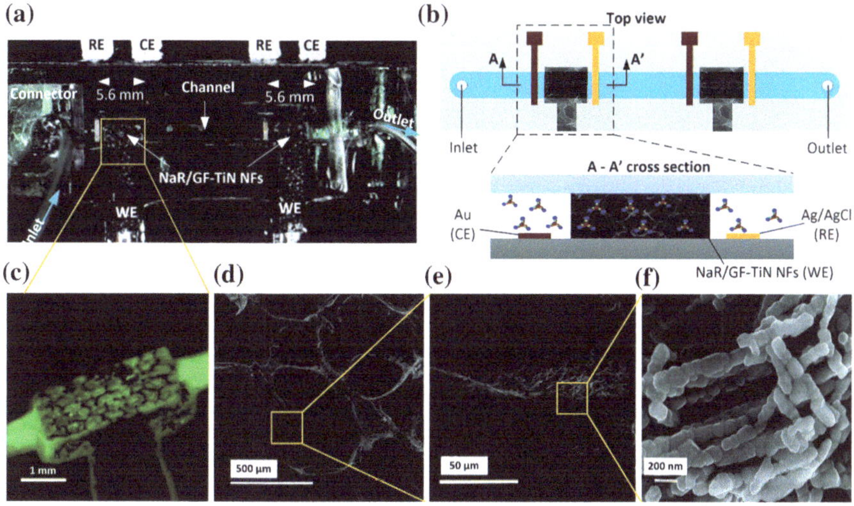

Fig. 4.8 **a** Photograph of the microfluidic device used to detect nitrate ions. **b** Schematic diagram of the sensor using embedded GO–TiN nanofibers used as a working electrode. **c** Fluorescence micrograph shows in situ integration of a GO–TiN nanofiber-based electrode into a microchannel with a liquid-phase photopolymerization. **d** FESEM micrograph of the GF–TiN nanofiber composite, **e** details of the GF scaffolds modified by TiN nanofibers, and **f** a magnified image of TiN nanofibers. Reproduced with permission from Ali et al. (2017)

2009; Kiran and Sampath 2013). Recently, Ali et al. (2017) have developed soil nutrient sensors via in situ incorporation of graphene foam–titanium nitride (TiN) nanofiber-based bioscaffolds and microfluidic assemblies. Figure 4.8 illustrates the microfluidic device for the detection of nitrate ions along with the nanostructured semiconductor/graphene nanoassembly. In another work, the same group (Ali et al. 2016b) have described a label-free microfluidic immunosensor which has femtomolar sensitivity and high selectivity for early stage of epidermal growth factor receptor 2 (EGFR2 or ErbB2) protein detection. The sensor uses a distinctively structured immunoelectrode fabricated out of hierarchical porous graphene foam improved with carbon-doped TiO_2 nanofibers which is shown in Fig. 4.9.

In addition to those application discussed above, photocatalyst/carbon nanocomposites have exposed a broad range of new applications in stem cell research (Bressan et al. 2014; El-Sadik 2010), biomaterial sciences (Park et al. 2014), and biochemistry (Cui et al. 2013) owing to their promising morphology and ideal physicochemical properties.

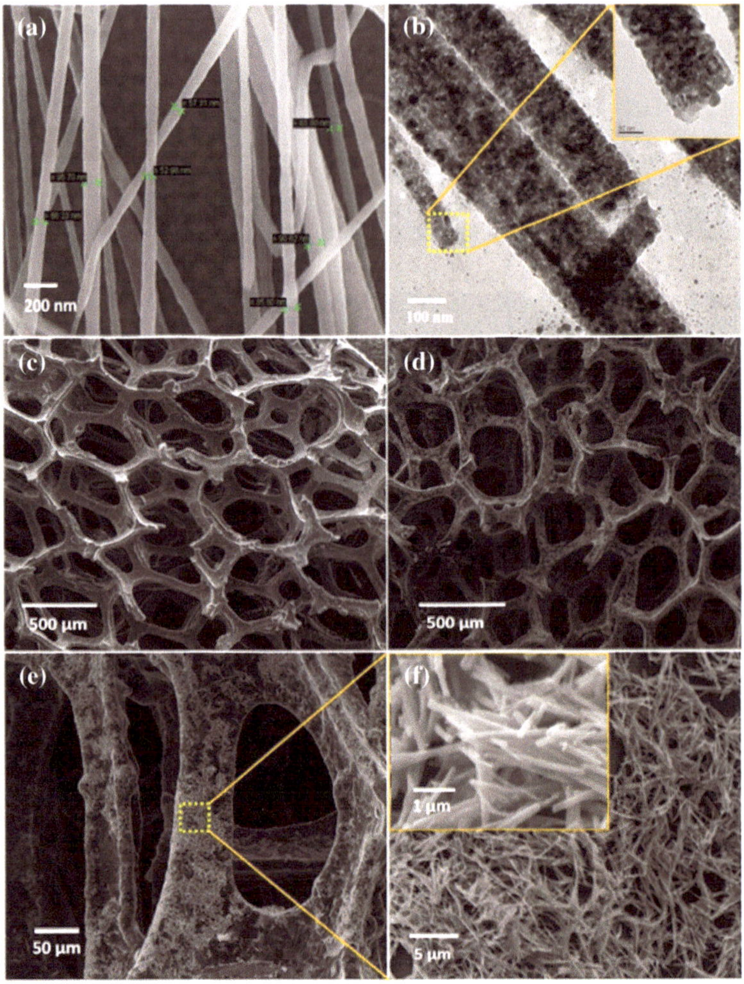

Fig. 4.9 **a** FESEM and **b** TEM micrograph of prepared carbon-doped TiO_2 nanofibers. Inset in '**b**' displays a magnified view of a single nanofiber. **c** FESEM micrograph of 3D graphene foam, **d–f** FESEM micrograph of graphene foam–TiO_2 nanofiber composite. The inset in '**f**' shows a magnified view of the composite material. Reprinted with permission from Ali et al. (2016b)

4.5 Recycling of Photocatalyst

Managing waste materials is one of the utmost combative issues faced by many countries. Discarding of waste materials in the open environment is more regulated now by many industrial countries. Thereby, the spent catalyst recycling is very important. It is also advantageous to refineries, power plants, cement plants, and other processes which use and produce by-products with chemical and mineral value. However, non-reusable photocatalytic particles significantly escalate the operational cost that considerably limits their actual applications. Thus, the spent photocatalyst recycling is indispensable for wastewater treatment plants, as this can decrease the cost of the whole waste management. Therefore, an effective work is essential for developing photocatalysts that retain nearly same photoactivity during each cycle and could be easily reused many times. Correspondingly, it is very important to design photocatalysts that are stress-free to separate in order not to lose any material value in waste stream. In addition to recyclability, photostability is an important measure when selecting a suitable photocatalyst material. In the literature, very few papers have been worried about the reusable property of the photocatalysts (Mondal et al. 2013, 2014; Gupta et al. 2015; Cong et al. 2007; Sun et al. 2008; Nishijima et al. 2007).

Photocatalysts have distinctive properties and benefits/drawbacks. However, the disadvantages of photocatalysts generally contain the necessity of an enormous amount of catalyst particles, trouble in recycling the catalyst, and problematical aggregation of nanoparticles that turn into large particles (Acar et al. 2016; Costa and Prado 2009). The powder of nanoparticle catalysts frequently undergoes deactivation, suffers from aggregation, and becomes difficult in settling for recycling after the early run. A recycling plan built on interchangeable water is very competent. Photocatalytic activities in aqueous as well as in organic media allow an easy retrieval of homogeneous catalyst particles from their product.

The other clever option is the membrane filtration technique for recycling soluble photocatalysts. Ultra- and nanofiltration are well-known methods for recycling photocatalysts due to their credible use in organic, enzymatic, and homogeneous catalysis-related uses at the laboratory scale as well as large industries (Taha 2015). In this context of recycling, magnetic nanoparticle photocatalysts are very convenient since they can be recycled easily with the applied external magnetic field (Dominguez et al. 2017). For example, Zhang et al. (2013b) prepared magnetic $BiFeO_3$ nanofibers by electrospinning for under visible light photocatalysis. Also, these fibrous photocatalytic nanostructures exhibit robust and active magnetic performance, which made their recycling and reuse very easy. Santala et al. (2009) recommended the use of an external magnetic field for recovery and recycling of electrospun magnetic nanofiber composite. Recently, Tarigh et al. (2015) fabricated magnetic multiwalled carbon nanotube–TiO_2 nanocomposite by electrostatic adsorption for the use in enhanced photodegradation of malachite green. They have shown that the composites can be recycled and reused for degradation by simply washing with ethanol three times.

4.6 Conclusions and Future Scopes

The chapter discusses the recent updates on the synthesis of nanocomposites containing carbon nanostructures and metal-oxide semiconductors, for example, TiO_2, ZnO, ZrO_2, SnO_2, WO_3 used as photocatalyst materials. The UV-, visible-, and solar-mediated photocatalytic activity of the carbon-based nanocomposites is advanced than that of the pure metal oxide or mechanical mixture of the metal oxide and carbon nanoparticles.

After reviewing more than hundreds of the recent publications in the field of photocatalysis on the character of several carbon–semiconductor nanocomposite photocatalyst particles and their synthesis and applications toward water treatment, it is evident that there are variety of composite photocatalyst nanostructures available based on the methods of fabrication and the area of their applications. Among them, quantum dots, carbon nanospheres, carbon nanotubes, graphene, and carbon nanofibers with semiconductor photocatalytic morphologies are steady and retain their physical structure when reused in wastewater treatment, bacterial cleansing, and air decontamination under ultraviolet, visible light photocatalysis. These composite photocatalysts are considered good support materials for semiconductor photocatalysts because those carbon nanostructures provide an easily accessible huge surface area support with high value active sites. Similarly, they help in stabilizing charge carrier separation by trapping electrons, thus delaying electron–hole recombination process by modifying bandgap and sensitization.

The key issue with photocatalytic oxidation falls in the high cost of reagents (such as hydrogen peroxide) or the source of energy (e.g., ultraviolet light source) and the high value of bandgap energy of quite a lot of photocatalysts. For the photocatalysis, the cost of the semiconductors and energy demand could be significantly condensed using solar energy and suitable nanomorphology which allows an accessible ultra-high active surface area. Altering their optical, physicochemical, and electronic characteristics, it can be guaranteed that these composite photocatalysts proficiently reduce wastewater in visible light.

The synergistic effect, as represented in Table 4.1, showed the benefits of the application of the photocatalysis. Photocatalytic oxidation can be established as a competent alternate way for the oxidation and removal of harmful organic compounds under precise environments for managing industrial and agricultural contaminants present in wastewater.

Acknowledgements The authors would like to acknowledge Professor Ashutosh Sharma (2015 to till date), the Honorable Secretary to the Department of Science and Technology, for his constant support and inspiration to many generations of colloid and surface scientists and engineers and enormous contribution to the nanoscience and nanotechnology of India and to the world.

References

Acar C, Dincer I, Naterer GF (2016) Review of photocatalytic water-splitting methods for sustainable hydrogen production: review: photocatalysis for sustainable hydrogen. Int J Energy Res 40:1449–1473. doi:10.1002/er.3549

Adamu H, McCue AJ, Taylor RSF et al (2017) Simultaneous photocatalytic removal of nitrate and oxalic acid over Cu_2O/TiO_2 and Cu_2O/TiO_2–AC composites. Appl Catal B Environ 217:181–191. doi:10.1016/j.apcatb.2017.05.091

Akhavan O (2010) Graphene nanomesh by ZnO nanorod photocatalysts. ACS Nano 4:4174–4180. doi:10.1021/nn1007429

Akir S, Hamdi A, Addad A et al (2017) Facile synthesis of carbon-ZnO nanocomposite with enhanced visible light photocatalytic performance. Appl Surf Sci 400:461–470. doi:10.1016/j.apsusc.2016.12.212

Ali MA, Singh C, Mondal K et al (2016a) Mesoporous few-layer graphene platform for affinity biosensing application. ACS Appl Mater Interfaces 8:7646–7656. doi:10.1021/acsami.5b12460

Ali MA, Mondal K, Jiao Y et al (2016b) Microfluidic immuno-biochip for detection of breast cancer biomarkers using hierarchical composite of porous graphene and titanium dioxide nanofibers. ACS Appl Mater Interfaces 8:20570–20582. doi:10.1021/acsami.6b05648

Ali MA, Mondal K, Wang Y et al (2017) In situ integration of graphene foam–titanium nitride based bio-scaffolds and microfluidic structures for soil nutrient sensors. Lab Chip 17:274–285. doi:10.1039/C6LC01266C

Allen MJ, Tung VC, Kaner RB (2010) Honeycomb carbon: a review of graphene. Chem Rev 110:132–145. doi:10.1021/cr900070d

Ameta SC, Ameta R (2016) Solar energy conversion and storage: photochemical modes. CRC Press, Boca Raton, FL

Arshad A, Iqbal J, Mansoor Q, Ahmed I (2017) Graphene/SiO_2 nanocomposites: the enhancement of photocatalytic and biomedical activity of SiO_2 nanoparticles by graphene. J Appl Phys 121:244901. doi:10.1063/1.4979968

Asahi R (2001) Visible-light photocatalysis in nitrogen-doped titanium oxides. Science (80-) 293:269–271. doi:10.1126/science.1061051

Bailón-García E, Elmouwahidi A, Carrasco-Marín F et al (2017) Development of carbon-ZrO 2 composites with high performance as visible-light photocatalysts. Appl Catal B Environ 217:540–550. doi:10.1016/j.apcatb.2017.05.090

Bakos LP, Justh N, Hernádi K et al (2016) Core-shell carbon nanosphere-TiO_2 composite and hollow TiO_2 nanospheres prepared by atomic layer deposition. J Phys: Conf Ser 764:12005. doi:10.1088/1742-6596/764/1/012005

Basahel SN, Ali TT, Mokhtar M, Narasimharao K (2015) Influence of crystal structure of nanosized ZrO_2 on photocatalytic degradation of methyl orange. Nanoscale Res Lett. doi:10.1186/s11671-015-0780-z

Binas V, Venieri D, Kotzias D, Kiriakidis G (2017) Modified TiO_2 based photocatalysts for improved air and health quality. J Mater 3:3–16. doi:10.1016/j.jmat.2016.11.002

Blake DM, Maness P-C, Huang Z et al (1999) Application of the photocatalytic chemistry of titanium dioxide to disinfection and the killing of cancer cells. Sep Purif Methods 28:1–50. doi:10.1080/03602549909351643

Bora LV, Mewada RK (2017) Visible/solar light active photocatalysts for organic effluent treatment: fundamentals, mechanisms and parametric review. Renew Sustain Energy Rev 76:1393–1421. doi:10.1016/j.rser.2017.01.130

Bressan E, Ferroni L, Gardin C et al (2014) Graphene based scaffolds effects on stem cells commitment. J Transl Med. doi:10.1186/s12967-014-0296-9

Bustos-Ramírez K, Barrera-Díaz CE, De Icaza-Herrera M et al (2015) 4-chlorophenol removal from water using graphite and graphene oxides as photocatalysts. J Environ Heal Sci Eng. doi:10.1186/s40201-015-0184-0

Byrne JA, Eggins BR, Brown NMD et al (1998) Immobilisation of TiO_2 powder for the treatment of polluted water. Appl Catal B Environ 17:25–36. doi:10.1016/S0926-3373(97)00101-X

Cai R, Wu J, Sun L et al (2016) 3D graphene/ZnO composite with enhanced photocatalytic activity. Mater Des 90:839–844. doi:10.1016/j.matdes.2015.11.020

Cao L, Sahu S, Anilkumar P et al (2011) Carbon nanoparticles as visible-light photocatalysts for efficient CO_2 conversion and beyond. J Am Chem Soc 133:4754–4757. doi:10.1021/ja200804h

Chakrabarti S, Dutta B (2004) Photocatalytic degradation of model textile dyes in wastewater using ZnO as semiconductor catalyst. J Hazard Mater 112:269–278. doi:10.1016/j.jhazmat.2004.05.013

Chang G, Cheng Z, Warren R et al (2015) Highly efficient photocatalysts for surface hybridization of TiO_2 nanofibers with carbon films. Chem Plus Chem 80:827–831. doi:10.1002/cplu.201402427

Chen X, Liu L, Yu PY, Mao SS (2011) Increasing solar absorption for photocatalysis with black hydrogenated titanium dioxide nanocrystals. Science (80-)331:746–750. doi:10.1126/science.1200448

Chen CS, Liu TG, Lin LW et al (2013) Multi-walled carbon nanotube-supported metal-doped ZnO nanoparticles and their photocatalytic property. J Nanoparticle Res. doi:10.1007/s11051-012-1295-5

Chen G, Wu S, Hui L et al (2016) Assembling carbon quantum dots to a layered carbon for high-density supercapacitor electrodes. Sci Rep. doi:10.1038/srep19028

Chen F, An W, Liu L et al (2017) Highly efficient removal of bisphenol A by a three-dimensional graphene hydrogel-AgBr@rGO exhibiting adsorption/photocatalysis synergy. Appl Catal B Environ 217:65–80. doi:10.1016/j.apcatb.2017.05.078

Cheng Y-F, Cao Q, Zhang J et al (2017a) Efficient photodegradation of dye pollutants using a novel plasmonic AgCl microrods array and photo-optimized surface-enhanced Raman scattering. Appl Catal B Environ 217:37–47. doi:10.1016/j.apcatb.2017.05.021

Cheng J, Wang Y, Xing Y et al (2017b) A stable and highly efficient visible-light photocatalyst of TiO_2 and heterogeneous carbon core–shell nanofibers. RSC Adv 7:15330–15336. doi:10.1039/C7RA00546F

Cong Y, Zhang J, Chen F, Anpo M (2007) Synthesis and characterization of nitrogen-doped TiO_2 nanophotocatalyst with high visible light activity. J Phys Chem C 111:6976–6982. doi:10.1021/jp0685030

Costa LL, Prado AGS (2009) TiO_2 nanotubes as recyclable catalyst for efficient photocatalytic degradation of indigo carmine dye. J Photochem Photobiol A Chem 201:45–49. doi:10.1016/j.jphotochem.2008.09.014

Cui G, Wang W, Ma M et al (2013) Rational design of carbon and TiO_2 assembly materials: covered or strewn, which is better for photocatalysis? Chem Commun 49:6415. doi:10.1039/c3cc42500b

Czech B, Buda W (2015) Photocatalytic treatment of pharmaceutical wastewater using new multiwall-carbon nanotubes/TiO_2/SiO_2 nanocomposites. Environ Res 137:176–184. doi:10.1016/j.envres.2014.12.006

Daneshvar N, Salari D, Khataee A (2004) Photocatalytic degradation of azo dye acid red 14 in water on ZnO as an alternative catalyst to TiO_2. J Photochem Photobiol A Chem 162:317–322. doi:10.1016/S1010-6030(03)00378-2

Dang X, Zhang X, Lu Z et al (2014) Construction of Au@TiO_2/graphene nanocomposites with plasmonic effect and super adsorption ability for enhanced visible-light-driven photocatalytic organic pollutant degradation. J Nanoparticle Res. doi:10.1007/s11051-013-2215-z

Das RK, Kar JP, Mohapatra S (2016) Enhanced photodegradation of organic pollutants by carbon quantum dot (CQD) deposited Fe_3O_4@mTiO_2 Nano-Pom-Pom Balls. Ind Eng Chem Res 55:5902–5910. doi:10.1021/acs.iecr.6b00792

Das S, Sinha S, Das B et al (2017) Disinfection of multidrug resistant escherichia coli by solar-photocatalysis using Fe-doped ZnO nanoparticles. Sci Rep. doi:10.1038/s41598-017-00173-0

Dey NK, Kim MJ, Kim K-D et al (2011) Adsorption and photocatalytic degradation of methylene blue over TiO_2 films on carbon fiber prepared by atomic layer deposition. J Mol Catal A: Chem 337:33–38. doi:10.1016/j.molcata.2011.01.010

Dindar B, Içli S (2001) Unusual photoreactivity of zinc oxide irradiated by concentrated sunlight. J Photochem Photobiol A Chem 140:263–268. doi:10.1016/S1010-6030(01)00414-2

Dominguez S, Huebra M, Han C et al (2017) Magnetically recoverable TiO_2-WO_3 photocatalyst to oxidize bisphenol A from model wastewater under simulated solar light. Environ Sci Pollut Res 24:12589–12598. doi:10.1007/s11356-016-7564-6

Dong S, Feng J, Fan M et al (2015) Recent developments in heterogeneous photocatalytic water treatment using visible light-responsive photocatalysts: a review. RSC Adv 5:14610–14630. doi:10.1039/C4RA13734E

Dunlop PSM, Byrne JA, Manga N, Eggins BR (2002) The photocatalytic removal of bacterial pollutants from drinking water. J Photochem Photobiol A Chem 148:355–363. doi:10.1016/S1010-6030(02)00063-1

Dwivedi N, Kumar S, Malik HK (2011) Nanostructured Titanium/Diamond-like carbon multilayer films: deposition, characterization, and applications. ACS Appl Mater Interfaces 3:4268–4278. doi:10.1021/am200939j

El-Sadik A (2010) Nanoparticle-labeled stem cells: a novel therapeutic vehicle. Clin Pharmacol Adv Appl 9. doi:10.2147/cpaa.s8931

Fan H, Zhao X, Yang J et al (2012) ZnO–graphene composite for photocatalytic degradation of methylene blue dye. Catal Commun 29:29–34. doi:10.1016/j.catcom.2012.09.013

Fang J, Wang W, Zhu C et al (2017) CdS/Pt photocatalytic activity boosted by high-energetic photons based on efficient triplet–triplet annihilation upconversion. Appl Catal B Environ 217:100–107. doi:10.1016/j.apcatb.2017.05.069

Fernando KAS, Sahu S, Liu Y et al (2015) Carbon quantum dots and applications in photocatalytic energy conversion. ACS Appl Mater Interfaces 7:8363–8376. doi:10.1021/acsami.5b00448

Fujishima A, Zhang X, Tryk D (2008) TiO_2 photocatalysis and related surface phenomena. Surf Sci Rep 63:515–582. doi:10.1016/j.surfrep.2008.10.001

Gaya UI, Abdullah AH (2008) Heterogeneous photocatalytic degradation of organic contaminants over titanium dioxide: a review of fundamentals, progress and problems. J Photochem Photobiol C Photochem Rev 9:1–12. doi:10.1016/j.jphotochemrev.2007.12.003

Goswami DY, Trivedi DM, Block SS (1997) Photocatalytic disinfection of indoor air. J Sol Energy Eng 119:92. doi:10.1115/1.2871871

Gu C, Xiong S, Zhong Z et al (2017) A promising carbon fiber-based photocatalyst with hierarchical structure for dye degradation. RSC Adv 7:22234–22242. doi:10.1039/C7RA02583A

Gumy D, Rincon AG, Hajdu R, Pulgarin C (2006) Solar photocatalysis for detoxification and disinfection of water: different types of suspended and fixed TiO_2 catalysts study. Sol Energy 80:1376–1381. doi:10.1016/j.solener.2005.04.026

Gupta A, Mondal K, Sharma A, Bhattacharya S (2015) Superhydrophobic polymethylsilsesquioxane pinned one dimensional ZnO nanostructures for water remediation through photo-catalysis. RSC Adv 5:45897–45907. doi:10.1039/C5RA02938D

Gutteridge JM (1995) Lipid peroxidation and antioxidants as biomarkers of tissue damage. Clin Chem 41:1819–1828

Herring NP, Almahoudi SH, Olson CR, El-Shall MS (2012) Enhanced photocatalytic activity of ZnO–graphene nanocomposites prepared by microwave synthesis. J Nanoparticle Res. doi:10.1007/s11051-012-1277-7

Hosseini SA, Babaei S (2016) Graphene oxide/zinc oxide (GO/ZnO) nanocomposite as a superior photocatalyst for degradation of methylene blue (MB)-process modeling by response surface methodology (RSM). J Braz Chem Soc. doi:10.5935/0103-5053.20160176

Hu L, Huo K, Chen R et al (2011) Recyclable and high-sensitivity electrochemical biosensing platform composed of carbon-doped TiO_2 nanotube arrays. Anal Chem 83:8138–8144. doi:10.1021/ac201639m

Hung-Low F, Ramirez DA, Peterson GR et al (2016) Development of a carbon-supported Sn–SnO_2 photocatalyst by a new hybridized sol–gel/dextran approach. RSC Adv 6:21019–21025. doi:10.1039/C6RA01129B

Ireland JC, Klostermann P, Rice EW, Clark RM (1993) Inactivation of *Escherichia coli* by titanium dioxide photocatalytic oxidation. Appl Environ Microbiol 59:1668–1670

Jafari T, Moharreri E, Amin A et al (2016) Photocatalytic water splitting—the untamed dream: a review of recent advances. Molecules 21:900. doi:10.3390/molecules21070900

Ji X, Lou W, Wang Q et al (2012) Sol-gel-derived hydroxyapatite-carbon nanotube/titania coatings on Titanium substrates. Int J Mol Sci 13:5242–5253. doi:10.3390/ijms13045242

Ji Z, Shen X, Li M et al (2013) Synthesis of reduced graphene oxide/CeO_2 nanocomposites and their photocatalytic properties. Nanotechnology 24:115603. doi:10.1088/0957-4484/24/11/115603

Jiang T, Zhang L, Ji M et al (2013) Carbon nanotubes/TiO2 nanotubes composite photocatalysts for efficient degradation of methyl orange dye. Particuology 11:737–742. doi:10.1016/j.partic.2012.07.008

Jiang L, Yuan X, Pan Y et al (2017) Doping of graphitic carbon nitride for photocatalysis: a reveiw. Appl Catal B Environ 217:388–406. doi:10.1016/j.apcatb.2017.06.003

Jin Y, Wu M, Zhao G, Li M (2011) Photocatalysis-enhanced electrosorption process for degradation of high-concentration dye wastewater on TiO_2/carbon aerogel. Chem Eng J 168:1248–1255. doi:10.1016/j.cej.2011.02.026

Jo W-K, Lee J, Chun H-H (2014) Titania nanotubes grown on carbon fibers for photocatalytic decomposition of gas-phase aromatic pollutants. Mat (Basel) 7:1801–1813. doi:10.3390/ma7031801

Kalt P, Birzer C, Evans H et al (2014) A solar disinfection water treatment system for remote communities. Procedia Eng 78:250–258. doi:10.1016/j.proeng.2014.07.064

Kanakaraju D, Glass BD, Oelgemöller M (2014) Titanium dioxide photocatalysis for pharmaceutical wastewater treatment. Environ Chem Lett 12:27–47. doi:10.1007/s10311-013-0428-0

Katiyar S, Mondal K, Sharma A (2016) One-step sol–gel synthesis of hierarchically porous, flow-through carbon/silica monoliths. RSC Adv 6:12298–12310. doi:10.1039/C5RA26503G

Khan Z, Chetia TR, Vardhaman AK et al (2012) Visible light assisted photocatalytic hydrogen generation and organic dye degradation by CdS–metal oxide hybrids in presence of graphene oxide. RSC Adv 2:12122. doi:10.1039/c2ra21596a

Kim CH, Kim B-H, Yang KS (2011) Visible light-induced photocatalytic activity of Ag-containing TiO_2/carbon nanofibers composites. Synth Met 161:1068–1072. doi:10.1016/j.synthmet.2011.03.017

Kiran V, Sampath S (2013) Facile synthesis of carbon doped TiO_2 nanowires without an external carbon source and their opto-electronic properties. Nanoscale 5:10646. doi:10.1039/c3nr03309k

Krishnamoorthy K, Mohan R, Kim S-J (2011) Graphene oxide as a photocatalytic material. Appl Phys Lett 98:244101. doi:10.1063/1.3599453

Kumar VB, Borenstein A, Markovsky B et al (2016) Activated carbon modified with carbon nanodots as novel electrode material for supercapacitors. J Phys Chem C 120:13406–13413. doi:10.1021/acs.jpcc.6b04045

Lee WJ, Lee JM, Kochuveedu ST et al (2012) Biomineralized N-doped CNT/TiO_2 core/shell nanowires for visible light photocatalysis. ACS Nano 6:935–943. doi:10.1021/nn204504h

Lee HJ, Kim J, Park CH (2014) Fabrication of self-cleaning textiles by TiO_2-carbon nanotube treatment. Text Res J 84:267–278. doi:10.1177/0040517513494258

Li J, Wu N (2015) Semiconductor-based photocatalysts and photoelectrochemical cells for solar fuel generation: a review. Catal Sci Technol 5:1360–1384. doi:10.1039/C4CY00974F

Li Q, Xie R, Li YW et al (2007) Enhanced visible-light-induced photocatalytic disinfection of *E. coli* by carbon-sensitized nitrogen-doped titanium oxide. Environ Sci Technol 41:5050–5056. doi:10.1021/es062753c

Li X, Lin H, Chen X et al (2015a) Fabrication of TiO_2/porous carbon nanofibers with superior visible photocatalytic activity. New J Chem 39:7863–7872. doi:10.1039/C5NJ01189B

Li Y, Zhao M, Zhang N et al (2015b) Synthesis and photocatalytic activity of carbon spheres loaded Cu_2O/Cu composites. J Alloys Compd 643:106–110. doi:10.1016/j.jallcom.2015.03.254

Li J, Liu X, Piao X et al (2015c) Novel carbon sphere@Bi_2MoO_6 core–shell structure for efficient visible light photocatalysis. RSC Adv 5:16592–16597. doi:10.1039/C4RA16777E

Liang Y, Wu D, Fu R (2013) Carbon microfibers with hierarchical porous structure from electrospun fiber-like natural biopolymer. Sci Rep. doi:10.1038/srep01119

Licht S (2001) Multiple band gap semiconductor/electrolyte solar energy conversion. J Phys Chem B 105:6281–6294. doi:10.1021/jp010552j

Lin X, Rong F, Ji X, Fu D (2011) Carbon-doped mesoporous TiO_2 film and its photocatalytic activity. Microporous Mesoporous Mater 142:276–281. doi:10.1016/j.micromeso.2010.12.010

Lin W, Yu X, Shen Y et al (2017) Carbon dots/BiOCl films with enhanced visible light photocatalytic performance. J Nanoparticle Res. doi:10.1007/s11051-017-3764-3

Linic S, Christopher P, Ingram DB (2011) Plasmonic-metal nanostructures for efficient conversion of solar to chemical energy. Nat Mater 10:911–921. doi:10.1038/nmat3151

Liu X, Pan L, Zhao Q et al (2012) UV-assisted photocatalytic synthesis of ZnO–reduced graphene oxide composites with enhanced photocatalytic activity in reduction of Cr(VI). Chem Eng J 183:238–243. doi:10.1016/j.cej.2011.12.068

Liu X, Pan L, Lv T, Sun Z (2013) Investigation of photocatalytic activities over ZnO–TiO_2—reduced graphene oxide composites synthesized via microwave-assisted reaction. J Colloid Interface Sci 394:441–444. doi:10.1016/j.jcis.2012.11.047

Luo Y, Luo J, Jiang J et al (2012) Seed-assisted synthesis of highly ordered TiO_2@α-Fe_2O_3 core/shell arrays on carbon textiles for lithium-ion battery applications. Energy Environ Sci 5:6559. doi:10.1039/c2ee03396h

Maness PC, Smolinski S, Blake DM et al (1999) Bactericidal activity of photocatalytic TiO(2) reaction: toward an understanding of its killing mechanism. Appl Environ Microbiol 65:4094–4098

Martins NCT, Ângelo J, Girão AV et al (2016) N-doped carbon quantum dots/TiO_2 composite with improved photocatalytic activity. Appl Catal B Environ 193:67–74. doi:10.1016/j.apcatb.2016.04.016

Masih D, Ma Y, Rohani S (2017) Graphitic C_3N_4 based noble-metal-free photocatalyst systems: a review. Appl Catal B Environ 206:556–588. doi:10.1016/j.apcatb.2017.01.061

Matsuda A, Sreekantan S, Krengvirat W (2013) Well-aligned TiO_2 nanotube arrays for energy-related applications under solar irradiation. J Asian Ceram Soc 1:203–219. doi:10.1016/j.jascer.2013.07.001

Matsunaga T, Tomoda R, Nakajima T, Wake H (1985) Photoelectrochemical sterilization of microbial cells by semiconductor powders. FEMS Microbiol Lett 29:211–214. doi:10.1111/j.1574-6968.1985.tb00864.x

Matsunaga T, Tomoda R, Nakajima T et al (1988) Continuous-sterilization system that uses photosemiconductor powders. Appl Environ Microbiol 54:1330–1333

McAllister MJ, Li J-L, Adamson DH et al (2007) Single sheet functionalized graphene by oxidation and thermal expansion of graphite. Chem Mater 19:4396–4404. doi:10.1021/cm0630800

Mills A, Le Hunte S (1997) An overview of semiconductor photocatalysis. J Photochem Photobiol A Chem 108:1–35. doi:10.1016/S1010-6030(97)00118-4

Miura T, Tao R, Shibata S et al (2016) Geometries, electronic couplings, and hole dissociation dynamics of photoinduced electron-hole pairs in polyhexylthiophene-fullerene dyads rigidly linked by oligophenylenes. J Am Chem Soc 138:5879–5885. doi:10.1021/jacs.5b13414

Miyazaki H, Matsui H, Nagano T et al (2008) Synthesis and electronic behaviors of TiO_2/carbon clusters/Cr_2O_3 composite materials. Appl Surf Sci 254:7365–7369. doi:10.1016/j.apsusc.2008.05.332

Mohapatra L, Parida K (2017) A review of solar and visible light active oxo-bridged materials for energy and environment. Catal Sci Technol 7:2153–2164. doi:10.1039/C7CY00116A

Mondal K (2017) Recent advances in the synthesis of metal oxide nanofibers and their environmental remediation applications. Inventions 2:9. doi:10.3390/inventions2020009

Mondal K, Moitra P (2017) Metal-semiconductor core–shell nanostructured photocatalysts for environmental applications and their recycling process. In: Metal semiconductor core-shell nanostructures energy environmental application. Elsevier, pp 133–157

Mondal K, Kumar J, Sharma A (2013) TiO_2-nanoparticles-impregnated photocatalytic macroporous carbon films by spin coating. Nanomater Energy 2:121–133. doi:10.1680/nme.12.00034

Mondal K, Bhattacharyya S, Sharma A (2014) Photocatalytic degradation of naphthalene by electrospun mesoporous carbon-doped anatase TiO_2 nanofiber mats. Ind Eng Chem Res 53:18900–18909. doi:10.1021/ie5025744

Mondal K, Kumar R, Sharma A (2016) Metal-oxide decorated multilayered three-dimensional (3D) porous carbon thin films for supercapacitor electrodes. Ind Eng Chem Res 55:12569–12581. doi:10.1021/acs.iecr.6b03396

Mondal K, Ali MA, Singh C et al (2017) Highly sensitive porous carbon and metal/carbon conducting nanofiber based enzymatic biosensors for triglyceride detection. Sens Actuators B Chem 246:202–214. doi:10.1016/j.snb.2017.02.050

Mungondori HH, Tichagwa L (2012) Photo-catalytic activity of carbon/nitrogen doped TiO_2-SiO_2 under UV and visible light irradiation. Mater Sci Forum 734:226–236. doi:10.4028/www.scientific.net/MSF.734.226

Neimark AV, Ruetsch S, Kornev KG et al (2003) Hierarchical pore structure and wetting properties of single-wall carbon nanotube fibers. Nano Lett 3:419–423. doi:10.1021/nl034013x

Nguyen DCT, Cho KY, Oh W-C (2017) Synthesis of mesoporous SiO_2/Cu_2O–Graphene nanocomposites and their highly efficient photocatalytic performance for dye pollutants. RSC Adv 7:29284–29294. doi:10.1039/C7RA03526H

Nishijima K, Ohtani B, Yan X et al (2007) Incident light dependence for photocatalytic degradation of acetaldehyde and acetic acid on S-doped and N-doped TiO_2 photocatalysts. Chem Phys 339:64–72. doi:10.1016/j.chemphys.2007.06.014

Nishikiori H, Fujiwara S, Miyagawa S et al (2017) Crystal growth of titania by photocatalytic reaction. Appl Catal B Environ 217:241–246. doi:10.1016/j.apcatb.2017.05.076

Niu P, Zhang L, Liu G, Cheng H-M (2012) Graphene-like carbon nitride nanosheets for improved photocatalytic activities. Adv Funct Mater 22:4763–4770. doi:10.1002/adfm.201200922

Nosaka Y, Ishizuka Y, Miyama H (1986) Separation mechanism of a photoinduced electron-hole pair in metal-loaded semiconductor powders. Berichte der Bunsengesellschaft für Phys Chemie 90:1199–1204. doi:10.1002/bbpc.19860901216

Nyamukamba P, Tichagwa L, Mamphweli S, Petrik L (2017) Silver/carbon codoped titanium dioxide photocatalyst for improved dye degradation under visible light. Int J Photoenergy 2017:1–9. doi:10.1155/2017/3079276

Ohno T, Akiyoshi M, Umebayashi T et al (2004) Preparation of S-doped TiO_2 photocatalysts and their photocatalytic activities under visible light. Appl Catal A Gen 265:115–121. doi:10.1016/j.apcata.2004.01.007

Omar FS, Nay Ming H, Hafiz SM, Ngee LH (2014) Microwave synthesis of zinc oxide/reduced graphene oxide hybrid for adsorption-photocatalysis application. Int J Photoenergy 2014:1–8. doi:10.1155/2014/176835

Ouyang K, Dai K, Walker SL et al (2016) Efficient photocatalytic disinfection of *Escherichia coli* O157:H7 using C70-TiO_2 hybrid under visible light irradiation. Sci Rep. doi:10.1038/srep25702

Oza G, Pandey S, Gupta A et al (2013) Photocatalysis-assisted water filtration: using TiO_2-coated vertically aligned multi-walled carbon nanotube array for removal of Escherichia coli O157:H7. Mater Sci Eng, C 33:4392–4400. doi:10.1016/j.msec.2013.06.039

Pan D, Jiao J, Li Z et al (2015) Efficient separation of electron-hole pairs in graphene quantum dots by TiO_2 heterojunctions for dye degradation. ACS Sustain Chem Eng 3:2405–2413. doi:10.1021/acssuschemeng.5b00771

Pan Y-X, You Y, Xin S et al (2017) Photocatalytic CO_2 reduction by carbon-coated indium-oxide nanobelts. J Am Chem Soc 139:4123–4129. doi:10.1021/jacs.7b00266

Park J (2017) Visible and near infrared light active photocatalysis based on conjugated polymers. J Ind Eng Chem 51:27–43. doi:10.1016/j.jiec.2017.03.022

Park JE, Park IS, Bae TS, Lee MH (2014) Electrophoretic deposition of carbon nanotubes over TiO_2 nanotubes: evaluation of surface properties and biocompatibility. Bioinorg Chem Appl 2014:1–7. doi:10.1155/2014/236521

Pelaez M, Nolan NT, Pillai SC et al (2012) A review on the visible light active titanium dioxide photocatalysts for environmental applications. Appl Catal B Environ 125:331–349. doi:10.1016/j.apcatb.2012.05.036

Peraltahernandez J, Manriquez J, Measvong Y et al (2007) Photocatalytic properties of nano-structured TiO_2-carbon films obtained by means of electrophoretic deposition. J Hazard Mater 147:588–593. doi:10.1016/j.jhazmat.2007.01.053

Permadani I, Phasa DA, Pratiwi AW, Rahmawati F (2016) The composite of ZrO_2–TiO_2 produced from local zircon sand used as a photocatalyst for the degradation of methylene blue in a single batik dye wastewater. Bull Chem React Eng Catal 11:133. doi:10.9767/bcrec.11.2.539.133-139

Prakash K, Senthil Kumar P, Pandiaraj S et al (2016) Controllable synthesis of SnO_2 photocatalyst with superior photocatalytic activity for the degradation of methylene blue dye solution. J Exp Nanosci 11:1138–1155. doi:10.1080/17458080.2016.1188222

Rahimi R, Zargari S, Yousefi A et al (2015) Visible light photocatalytic disinfection of *E. coli* with TiO_2–graphene nanocomposite sensitized with tetrakis(4-carboxyphenyl)porphyrin. Appl Surf Sci 355:1098–1106. doi:10.1016/j.apsusc.2015.07.115

Rao CNR, Sood AK, Subrahmanyam KS, Govindaraj A (2009) Graphene: the new two-dimensional nanomaterial. Angew Chemie Int Ed 48:7752–7777. doi:10.1002/anie.200901678

Rashid J, Barakat MA, Ruzmanova Y, Chianese A (2015) $Fe_3O_4/SiO_2/TiO_2$ nanoparticles for photocatalytic degradation of 2-chlorophenol in simulated wastewater. Environ Sci Pollut Res 22:3149–3157. doi:10.1007/s11356-014-3598-9

Ren Y, Ma Z, Morris RE et al (2013) A solid with a hierarchical tetramodal micro-meso-macro pore size distribution. Nat Commun. doi:10.1038/ncomms3015

Réti B, Kiss GI, Gyulavári T et al (2017) Carbon sphere templates for TiO_2 hollow structures: preparation, characterization and photocatalytic activity. Catal Today 284:160–168. doi:10.1016/j.cattod.2016.11.038

Rey A, García-Muñoz P, Hernández-Alonso MD et al (2014) WO_3–TiO_2 based catalysts for the simulated solar radiation assisted photocatalytic ozonation of emerging contaminants in a municipal wastewater treatment plant effluent. Appl Catal B Environ 154–155:274–284. doi:10.1016/j.apcatb.2014.02.035

Roy P, Periasamy AP, Liang C-T, Chang H-T (2013) Synthesis of graphene-ZnO-Au nanocomposites for efficient photocatalytic reduction of nitrobenzene. Environ Sci Technol 47:6688–6695. doi:10.1021/es400422k

Rtimi S, Kiwi J (2017) Bactericide effects of transparent polyethylene photocatalytic films coated by oxides under visible light. Appl Catal B Environ 213:62–73. doi:10.1016/j.apcatb.2017.05.004

Saito T, Iwase T, Horie J, Morioka T (1992) Mode of photocatalytic bactericidal action of powdered semiconductor TiO_2 on mutans streptococci. J Photochem Photobiol B Biol 14:369–379. doi:10.1016/1011-1344(92)85115-B

Sakthivel S, Neppolian B, Shankar MV et al (2003) Solar photocatalytic degradation of azo dye: comparison of photocatalytic efficiency of ZnO and TiO_2. Sol Energy Mater Sol Cells 77:65–82. doi:10.1016/S0927-0248(02)00255-6

Salarian M, Xu WZ, Wang Z et al (2014) Hydroxyapatite–TiO $_2$ -based nanocomposites synthesized in supercritical CO_2 for bone tissue engineering: physical and mechanical properties. ACS Appl Mater Interfaces 6:16918–16931. doi:10.1021/am5044888

Saleh AT (2013) The role of carbon nanotubes in enhancement of photocatalysis. In: Synthesis of carbon nanotube/catalyst composites

Salih FM (2002) Enhancement of solar inactivation of *Escherichia coli* by titanium dioxide photocatalytic oxidation. J Appl Microbiol 92:920–926. doi:10.1046/j.1365-2672.2002.01601.x

Sampaio MJ, Silva CG, Marques RRN et al (2011) Carbon nanotube–TiO_2 thin films for photocatalytic applications. Catal Today 161:91–96. doi:10.1016/j.cattod.2010.11.081

Santala E, Kemell M, Leskelä M, Ritala M (2009) The preparation of reusable magnetic and photocatalytic composite nanofibers by electrospinning and atomic layer deposition. Nanotechnology 20:35602. doi:10.1088/0957-4484/20/3/035602

Sariciftci NS, Kraabel B, Lee CH et al (1994) Absence of photoinduced electron transfer from the excitonic electron-hole bound state in polydiacetylene conjugated polymers. Phys Rev B 50:12044–12051. doi:10.1103/PhysRevB.50.12044

Sawant VJ, Bamane SR, Kanase DG et al (2016) Encapsulation of curcumin over carbon dot coated TiO_2 nanoparticles for pH sensitive enhancement of anticancer and anti-psoriatic potential. RSC Adv 6:66745–66755. doi:10.1039/C6RA13851A

Schneider J, Matsuoka M, Takeuchi M et al (2014) Understanding TiO_2 photocatalysis: mechanisms and materials. Chem Rev 114:9919–9986. doi:10.1021/cr5001892

Sha D, Wang J, Wu X et al (2016) One-step and large-scale preparation of TiO_2/amorphous carbon composites with excellent visible light photocatalytic properties. RSC Adv 6:65607–65612. doi:10.1039/C6RA11250A

Shen Y, Wang W, Xiao K (2016) Synthesis of three-dimensional carbon felt supported TiO_2 monoliths for photocatalytic degradation of methyl orange. J Environ Chem Eng 4:1259–1266. doi:10.1016/j.jece.2016.01.026

Shi J, Zheng J, Wu P, Ji X (2008) Immobilization of TiO_2 films on activated carbon fiber and their photocatalytic degradation properties for dye compounds with different molecular size. Catal Commun 9:1846–1850. doi:10.1016/j.catcom.2008.02.018

Shi E, Zhang L, Li Z et al (2012) TiO_2-coated carbon nanotube-silicon solar cells with efficiency of 15%. Sci Rep. doi:10.1038/srep00884

Sies H (1985) Oxidative stress. Academic Press, London, Orlando

Singh P, Mondal K, Sharma A (2013) Reusable electrospun mesoporous ZnO nanofiber mats for photocatalytic degradation of polycyclic aromatic hydrocarbon dyes in wastewater. J Colloid Interface Sci 394:208–215. doi:10.1016/j.jcis.2012.12.006

Singh N, Mondal K, Misra M et al (2016) Quantum dot sensitized electrospun mesoporous titanium dioxide hollow nanofibers for photocatalytic applications. RSC Adv 6:48109–48119. doi:10.1039/C6RA04305D

Singh S, Mondal K, Sharma A (2017) ZnO nanoparticle fortified highly permeable carbon/silica monoliths as a flow-through media. Langmuir. doi:10.1021/acs.langmuir.7b01361

Stevenson M, Bullock K, Lin W-Y, Rajeshwar K (1997) Sonolytic enhancement of the bactericidal activity of irradiated titanium dioxide suspensions in water. Res Chem Intermed 23:311–323. doi:10.1163/156856797X00547

Su B-L, Vantomme A, Surahy L et al (2007) Hierarchical multimodal mesoporous carbon materials with parallel macrochannels. Chem Mater 19:3325–3333. doi:10.1021/cm070294o

Sun H, Bai Y, Jin W, Xu N (2008) Visible-light-driven TiO_2 catalysts doped with low-concentration nitrogen species. Sol Energy Mater Sol Cells 92:76–83. doi:10.1016/j.solmat.2007.09.003

Sunada K, Kikuchi Y, Hashimoto K, Fujishima A (1998) Bactericidal and detoxification effects of TiO_2 thin film photocatalysts. Environ Sci Technol 32:726–728. doi:10.1021/es970860o

Taha AA (2015) Direct synthesis of mesostructured carbon nanofibers decorated with silver-nanoparticles as a multifunctional membrane for water treatment. Adv Nat Sci Nanosci Nanotechnol 6:45003. doi:10.1088/2043-6262/6/4/045003

Tarigh GD, Shemirani F, Maz'hari NS (2015) Fabrication of a reusable magnetic multi-walled carbon nanotube–TiO_2 nanocomposite by electrostatic adsorption: enhanced photodegradation of malachite green. RSC Adv 5:35070–35079. doi:10.1039/c4ra15593a

Teng F, Zhang G, Wang Y et al (2014) The role of carbon in the photocatalytic reaction of carbon/TiO_2 photocatalysts. Appl Surf Sci 320:703–709. doi:10.1016/j.apsusc.2014.09.153

Thangavel S, Krishnamoorthy K, Krishnaswamy V et al (2015) Graphdiyne–ZnO nanohybrids as an advanced photocatalytic material. J Phys Chem C 119:22057–22065. doi:10.1021/acs.jpcc.5b06138

Tong H, Ouyang S, Bi Y et al (2012) Nano-photocatalytic materials: possibilities and challenges. Adv Mater 24:229–251. doi:10.1002/adma.201102752

Velasco LF, Haro M, Parmentier J et al (2013) Tuning the photocatalytic activity and optical properties of mesoporous TiO_2 spheres by a carbon scaffold. J Catal 2013:1–9. doi:10.1155/2013/178512

Wang R, Hashimoto K, Fujishima A et al (1998) Photogeneration of highly amphiphilic TiO_2 surfaces. Adv Mater 10:135–138. doi:10.1002/(SICI)1521-4095(199801)10:2<135:AID-ADMA135>3.0.CO;2-M

Wang Y, Di Y, Antonietti M et al (2010) Excellent Visible-light photocatalysis of fluorinated polymeric carbon nitride solids. Chem Mater 22:5119–5121. doi:10.1021/cm1019102

Wang W, Fang J, Shao S et al (2017a) Compact and uniform TiO_2@g-C_3N_4 core-shell quantum heterojunction for photocatalytic degradation of tetracycline antibiotics. Appl Catal B Environ 217:57–64. doi:10.1016/j.apcatb.2017.05.037

Wang J, Yang Z, Gao X et al (2017b) Core-shell g-C_3N_4@ZnO composites as photoanodes with double synergistic effects for enhanced visible-light photoelectrocatalytic activities. Appl Catal B Environ 217:169–180. doi:10.1016/j.apcatb.2017.05.034

Wei C, Lin WY, Zainal Z et al (1994) Bactericidal Activity of TiO_2 photocatalyst in aqueous media: toward a solar-assisted water disinfection system. Environ Sci Technol 28:934–938. doi:10.1021/es00054a027

Wei X, Jiang X, Wei J, Gao S (2016) Functional groups and pore size distribution do matter to hierarchically porous carbons as high-rate-performance supercapacitors. Chem Mater 28:445–458. doi:10.1021/acs.chemmater.5b02336

Williams G, Kamat PV (2009) Graphene−semiconductor nanocomposites: excited-state interactions between ZnO nanoparticles and graphene oxide †. Langmuir 25:13869–13873. doi:10.1021/la900905h

Williams G, Seger B, Kamat PV (2008) TiO_2–Graphene nanocomposites. UV-assisted photocatalytic reduction of graphene oxide. ACS Nano 2:1487–1491. doi:10.1021/nn800251f

Woan K, Pyrgiotakis G, Sigmund W (2009) Photocatalytic carbon-nanotube-TiO_2 composites. Adv Mater 21:2233–2239. doi:10.1002/adma.200802738

Wu X, Jaatinen E, Sarina S, Zhu HY (2017a) Direct photocatalysis of supported metal nanostructures for organic synthesis. J Phys D Appl Phys 50:283001. doi:10.1088/1361-6463/aa73f6

Wu H, Wu X-L, Wang Z-M et al (2017b) Anchoring titanium dioxide on carbon spheres for high-performance visible light photocatalysis. Appl Catal B Environ 207:255–266. doi:10.1016/j.apcatb.2017.02.027

Xing B, Shi C, Zhang C et al (2016) Preparation of TiO_2/activated carbon composites for photocatalytic degradation of RhB under UV light irradiation. J Nanomater 2016:1–10. doi:10.1155/2016/8393648

Xu H-Y, Wu L-C, Zhao H et al (2015) Synergic effect between adsorption and photocatalysis of metal-free g-C_3N_4 derived from different precursors. PLoS ONE 10:e0142616. doi:10.1371/journal.pone.0142616

Xu J, Cui Y, Han Y et al (2016) ZnO–graphene composites with high photocatalytic activities under visible light. RSC Adv 6:96778–96784. doi:10.1039/C6RA19622E

Yan H, McCreery RL (2009) Anomalous tunneling in carbon/alkane/TiO_2/gold molecular electronic junctions: energy level alignment at the metal/semiconductor interface. ACS Appl Mater Interfaces 1:443–451. doi:10.1021/am800126v

Yan SC, Li ZS, Zou ZG (2009) Photodegradation performance of g-C_3N_4 fabricated by directly heating melamine. Langmuir 25:10397–10401. doi:10.1021/la900923z

Yan SC, Li ZS, Zou ZG (2010) Photodegradation of rhodamine B and methyl orange over boron-doped g-C_3N_4 under visible light irradiation. Langmuir 26:3894–3901. doi:10.1021/la904023j

Yang J, Bai H, Tan X, Lian J (2006) IR and XPS investigation of visible-light photocatalysis—nitrogen–carbon-doped TiO_2 film. Appl Surf Sci 253:1988–1994. doi:10.1016/j.apsusc.2006.03.078

Yang M-Q, Zhang N, Xu Y-J (2013) Synthesis of fullerene– Carbon nanotube– and Graphene–TiO_2 nanocomposite photocatalysts for selective oxidation: a comparative study. ACS Appl Mater Interfaces 5:1156–1164. doi:10.1021/am3029798

Yang GD, Liu F, Wang L et al (2014) Selective deposition of nanocrystalline carbon films on GaN diodes in photocatalytic reactions. Cryst Eng Comm 16:10097–10102. doi:10.1039/C4CE01431F

Yao S, Song S, Shi Z (2014) Adsorption properties and photocatalytic activity of TiO_2/activated carbon fiber composite. Russ J Phys Chem A 88:1066–1070. doi:10.1134/S0036024414060296

Yi Q, Wang H, Cong S et al (2016) Self-cleaning glass of photocatalytic anatase TiO_2@carbon nanotubes thin film by polymer-assisted approach. Nanoscale Res Lett. doi:10.1186/s11671-016-1674-4

Yoon TP, Ischay MA, Du J (2010) Visible light photocatalysis as a greener approach to photochemical synthesis. Nat Chem 2:527–532. doi:10.1038/nchem.687

Yu JC, Ho W, Yu J et al (2005) Efficient visible-light-induced photocatalytic disinfection on sulfur-doped nanocrystalline Titania. Environ Sci Technol 39:1175–1179. doi:10.1021/es035374h

Zhai W, Li G, Yu P et al (2013) Silver phosphate/carbon nanotube-stabilized pickering emulsion for highly efficient photocatalysis. J Phys Chem C 117:15183–15191. doi:10.1021/jp404456a

Zhang H, Zong R, Zhu Y (2009) Photocorrosion inhibition and photoactivity enhancement for zinc oxide via hybridization with monolayer polyaniline. J Phys Chem C 113:4605–4611. doi:10.1021/jp810748u

Zhang D, Wu J, Zhou B et al (2013a) Efficient photocatalytic activity with carbon-doped SiO_2 nanoparticles. Nanoscale 5:6167. doi:10.1039/c3nr01314f

Zhang X, Liu H, Zheng B et al (2013b) Photocatalytic and magnetic behaviors observed in $BiFeO_3$ nanofibers by electrospinning. J Nanomater 2013:1–7. doi:10.1155/2013/917948

Zhang J, Li W, Li Y et al (2017) Self-optimizing bifunctional CdS/Cu_2S with coexistence of light-reduced Cu0 for highly efficient photocatalytic H_2 generation under visible-light irradiation. Appl Catal B Environ 217:30–36. doi:10.1016/j.apcatb.2017.05.074

Zhao W, Ma W, Chen C et al (2004) Efficient degradation of toxic organic pollutants with Ni_2O_3/TiO_{2-x} B_x under visible irradiation. J Am Chem Soc 126:4782–4783. doi:10.1021/ja0396753

Chapter 5
Dye Wastewater Treatment Using Carbonaceous Materials by Microwave-Assisted Methods

Remya Neelancherry

Abstract Dyes are extensively used in various industries such as textiles, paper, plastic, leather, food, and cosmetics. Carbonaceous-material-based adsorption is widely adopted method for dye wastewater treatment owing to its porous structure and surface activity. In recent years, microwave (MW)-assisted dye wastewater treatment gained interest owing to the specific heating properties of MW such as volumetric heating, rapid heating, and hot spot formation. Solid-phase heating of carbon by MW was used for the activation of carbonaceous material as well as for the regeneration of spent carbon. The activation or regeneration is done in presence acids or oxidants as activation agent. On the other hand, adsorption and simultaneous degradation of dye from wastewater were achieved in solvent-phase MW heating. In solvent-phase MW heating for dye wastewater treatment, the synergistic effect of adsorption on granular activated carbon (GAC) and degradation by MW is contributed to the high degradation efficiency.

Keywords Dye wastewater · Microwave · Activated carbon · Carbon activation Regeneration

5.1 Introduction

Dyes are natural or synthetic material used to add or change the color of a substance. They are extensively used in various industries such as textiles, paper, plastic, leather, food, and cosmetics. Despite of extensive use, dyes and its degradation products such as aromatic amines are highly toxic to cause carcinogenesis, mutagenesis, and respiratory diseases (Robinson et al. 2001). Therefore, the discharge of dye wastewater into the environment will pose serious

R. Neelancherry (✉)
School of Infrastructure, Indian Institute of Technology, Bhubaneswar 752050, Odisha, India
e-mail: remyamadhav@gmail.com

S. Bhattacharya et al. (eds.), *Water Remediation*, Energy, Environment, and Sustainability, https://doi.org/10.1007/978-981-10-7551-3_5

environmental, esthetical, and health problems. The treatment of wastewater containing dyes is challenging owing to the presence of complex aromatic compounds with high stability. Hence, the conventional primary and secondary systems employed for organic wastewater treatment seldom suit for these effluents. The methods such as coagulation, flocculation, biosorption, photo-decomposition, and ultrafiltration have seldom applied for the treatment of dye wastewater due to the relatively high operational costs and the low removal ratios (Lee et al. 1990; Mall and Upaphyay 1995). On the other hand, the commonly adopted methods such as adsorption using activated carbon and filtration cause generation of large amount of sludge and solid waste leading to high treatment cost (Park et al. 2010).

Incorporation of MW with the carbon compounds such as GAC is the recent development in the adsorptive removal/degradation of dye from wastewater. Microwave-assisted method for dye wastewater treatment has shown tremendous potential in the degradation of dyes and the regeneration of the adsorbent (Quan et al. 2004; Peng et al. 2004). The mechanism of MW heating and research progress in the dye wastewater treatment by MW-incorporated carbon compound treatment approach is summarized in this chapter.

5.2 Microwave (MW)

Microwave is a part of electromagnetic spectrum (Fig. 5.1) with frequency ranging from 300 MHz to 300 GHz (wavelength 1 m–1 cm). The domestic microwave ovens generally operate at a frequency of 2450 MHz, corresponding to a wavelength of 12.24 cm and with energy of 1.02×10^{-5} eV/MW photon (about 1 J/mole). On the other hand, industrial heating applications make use of a frequency of 915 MHz (wavelength of 32.7 cm) (Eskicioglu et al. 2007). Magnetron is the most

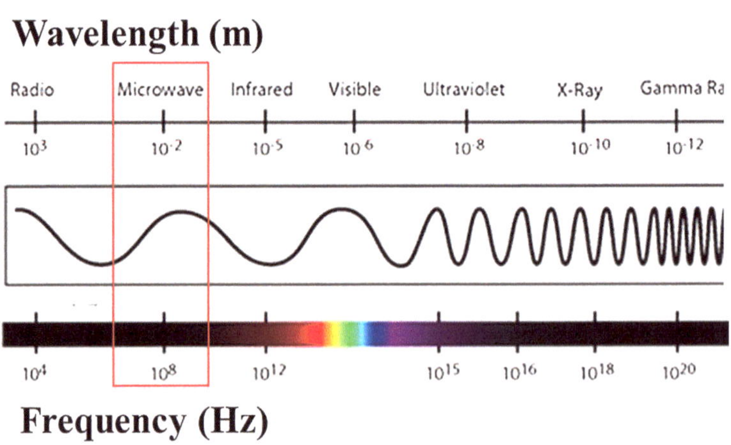

Fig. 5.1 Electromagnetic spectrum indicating the position of microwave

commonly used device for microwave generation owing to its reliability and cost efficiency compared to other devices such as klystrons, power grid tubes, traveling wave tubes, and gyrotrons (National Research Council (NRC) 1994).

5.3 MW Heating Mechanism

MW irradiation possesses several advantages over conventional heating such as rapid and volumetric heating, selective heating, increased reaction rates, reduced activation energies, and high energy efficiency (Remya and Lin 2011). Microwave heating results from the dielectric heating, conduction loss heating, and magnetic loss heating (Horikoshi and Serpone 2011). The thermal energy produced by MW depends on the strength of the electric field and magnetic fields of MW radiation ($|E|$ and $|H|$, respectively), frequency of the microwaves (f, GHz), electrical conductivity of the medium (σ), permittivity of the vacuum (ε^{o}, 8.854×10^{-12} F/m), the relative dielectric loss factor (ε''), magnetic permeability of the vacuum (μ', $4\pi \times 10^{-7}$ N/A^{-2}), and the relative magnetic loss (μ'').

The thermal energy P produced per unit volume originating from microwave radiation (W/m^3) is given in Eq. 5.1:

$$P = \pi f \varepsilon^{o} \varepsilon'' |E|^2 + \frac{1}{2}\sigma |E|^2 + \pi f \mu' \mu'' |H|^2 \tag{5.1}$$

The three terms in Eq. 5.1 denote the dielectric loss heating, conduction loss heating, and the magnetic loss heating, respectively. However, the dielectric loss heating and conduction loss heating which results from the interaction of materials with electric field are primarily responsible for MW heating of materials.

Dielectric loss heating, also referred as dipolar polarization, results from intermolecular inertia. When the dipole is subjected to a high-frequency alternating electric field of the MW, rotation (reversing) of the dipole cannot adequately follow the rate of change of direction of the electric field. This leads to a time delay in the rotation of dipole corresponding to the reversal of direction of electric field, resulting in a substantial quantity of energy to be spent which in turn turns into heat (Hidaka et al. 2007). The conduction loss heating occurs when an electrical conductor is irradiated with MW. On exposure to MW, the charge carriers (electrons, ions, etc.) move through the material under the influence of the electric field resulting in a polarization. Subsequently, the induced currents cause heating of the sample due to electrical resistance. Figures 5.2 and 5.3 illustrate the dielectric loss heating and conduction loss heating mechanism of MW, respectively.

Dielectric loss heating is mainly responsible for the MW heating in solvent media. Therefore, the dielectric properties play a critical role in the interaction between MW electric filed and solution. The ability of the material to convert electromagnetic energy into heat by dielectric loss heating is defined by the loss

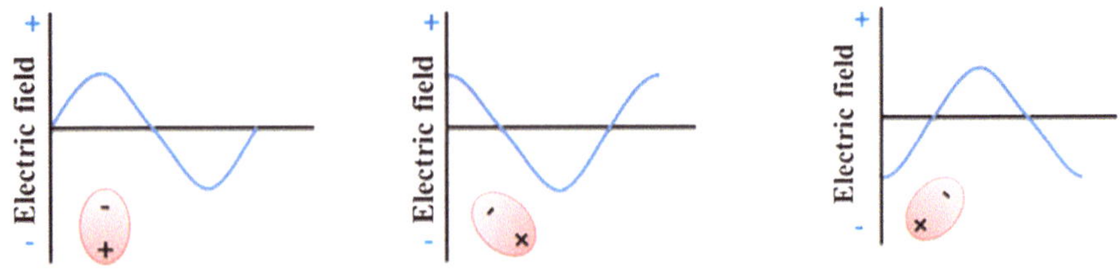

Fig. 5.2 Mechanism of dielectric loss heating on MW exposure

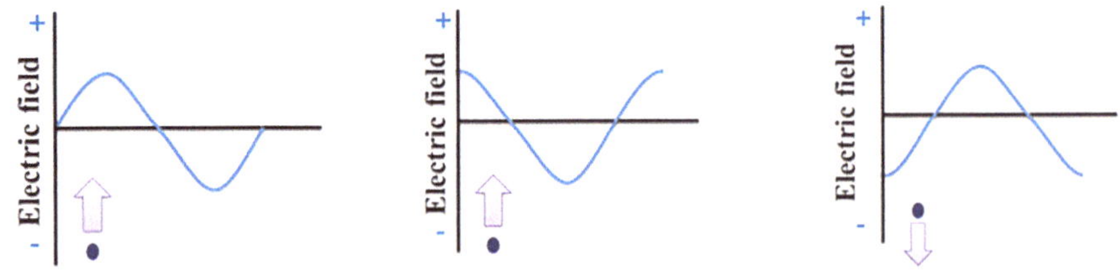

Fig. 5.3 MW heating mechanism by conduction loss heating

tangent (Eq. 5.2), which is the ratio of the dielectric loss factor ε'' to the dielectric constant ε'.

$$\tan \partial = \varepsilon''/\varepsilon' \quad (5.2)$$

The dielectric loss factor ε'' accounts for the loss energy-dissipative mechanisms in the material. On the other hand, the dielectric constant (ε') is the ratio permittivity of the material to the permittivity of the vacuum. It is the measure of the amount of energy from an external electric field stored in the material because of the polarization of material on exposure to MW. Therefore, a material with a high loss factor is easily heated by microwave. On the other hand, the material with a very low ε'' is transparent to microwave effect.

Most solvents including water and ethanol have permanent dipoles, which cause friction and heat. On the other hand, solid-phase heating of materials by microwave is a combined effect of dielectric heating and conduction loss heating, referred as interfacial polarization.

In the treatment of dye wastewater treatment using carbonaceous material, MW irradiation of the material (used as adsorbent/catalyst) prior to the wastewater treatment was employed to improve the adsorbent/catalytic property of the material. On the other hand, microwave was also employed in combination with carbonaceous material to achieve improved degradation of the dye from aqueous media. The former employed a solid-phase heating for of material, whereas the latter deploys the use of MW for solvent-phase heating.

5.4 Solid-Phase Microwave Heating for Enhancement in Adsorbent/Catalyst Property

In the solid-phase microwave heating, the effect of dielectric loss heating is negligible owing to few or no freely rotating dipoles. In its place, the interaction between microwaves and electrons plays an important role in MW heating of such systems.

5.4.1 MW for Activation of Carbon

Microwave for activation/regeneration of carbon is solid-phase heating, wherein the main targets are solid-state carbon particles. Microwave heating of carbon-based solids is caused by interfacial polarization (Maxwell–Wagner–Sillars polarization), which is a combination of dipolar polarization and ionic conduction (Kim et al. 2014). In addition, carbon-based solids possess semiconducting features because of delocalized π-electrons in the surface of the activated carbon which are free to move. MW irradiation causes increase in the kinetic energy of electrons on the surface of the activated carbon, which enables the electrons to jump out of the material. This would result in ionizing the surrounding atmosphere and formation of high temperature, called hot spot. The schematic representation of hot spot formation is shown in Fig. 5.4. These hot spots are plasmas, which are confined to a tiny region of space and last for fraction of a second. The temperature of hot spot can ordinarily reach 1200 °C (Jones et al. 2002). The high temperature on the carbon surface imparts surface modification of the carbon-containing compound to improve the micro-porosity. Figure 5.5 shows the improved micro-porosity of GAC surface on exposure to MW (Remya and Lin 2015).

Microwave thermal treatment showed enhanced activation of carbon and subsequent improvement in the adsorption capacity of dye from wastewater (Table 5.1). On the other hand, MW-assisted activation in presence of acid and

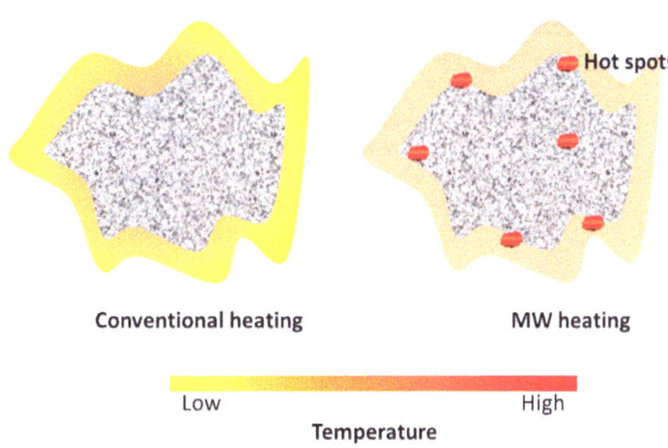

Fig. 5.4 Schematic representation of hot spot formation on carbon surface on MW irradiation

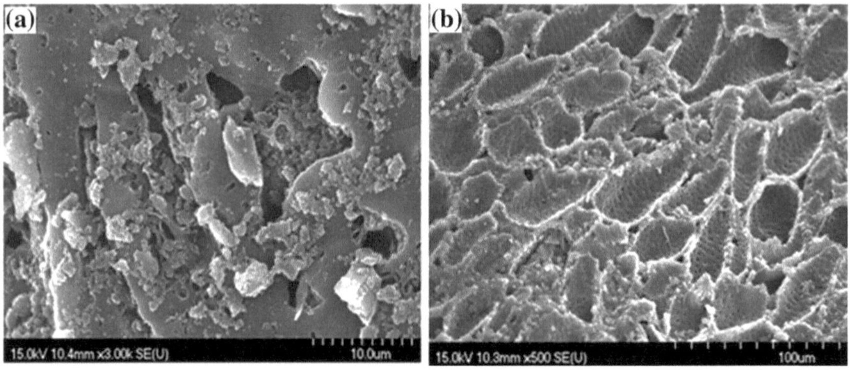

Fig. 5.5 SEM picture of GAC **a** before and **b** after exposure to MW (Remya and Lin 2015)

Table 5.1 Dye removal by adsorption using carbon activated by MW thermal treatment

S. No.	Carbon compound	Prepared adsorbent	Preparation method	Dye	Removal at optimum condition	Reference
1	Papaya leaf stem	Activated carbon	Mixing of dried and crushed papaya leaf stem with predetermined quantity of $ZnCl_2$ (30% w/v) solution and subsequent microwaving for 10 min	Malachite green (MG) Reactive orange (RO)	352 mg/g 163 mg/g	(Mohan Raj and Asaithambi 2014)
2	Epicarp of *Ricinus communis* (RC)	MW activated RC (MRC) and MW-assisted zinc chloride activated RC (ZRC)	Air-dried and powdered epicarp of *Ricinus communis* was dried in MW oven at fixed temperature and time	Malachite green (MG)	12.65 mg/g of MRC and 24.39 mg/g of ZRC	(Makeswari and Santhi 2013)
3	Biochar obtained as pyrolysis by-product of biomass	MW-H_2O_2-treated biochar (HC)	H_2O_2 treatment for 24 h followed by microwave pyrolysis at 600 W	Remazol brilliant blue (RBB) Methylene blue (MB)	83 mg/g 91 mg/g	(Nair and Vinu 2016)
4	Oil palm fiber (PF)	Palm fiber activated carbon (PFAC)	Air-dried PF was crushed and carbonized at 700 °C, and the product obtained was mixed with fixed quantity of KOH	MB	382.32 mg/g	(Foo and Hameed 2012)

(continued)

Table 5.1 (continued)

S. No.	Carbon compound	Prepared adsorbent	Preparation method	Dye	Removal at optimum condition	Reference
5	Wood sawdust (WS)	Wood sawdust activated carbon (WSAC)	Carbonation of air-dried and crushed WS was carbonized under N_2 atmosphere, followed by mixing of predetermined amount of K_2CO_3	MB	423.17 mg/g	(Foo and Hameed 2012)
6	Activated sludge	Activated carbon	Dewatered sludge (solid content 10% w/w) was mixed with acetic acid (1:1 by weight), and the mixture was heated in a microwave oven at 800 W for 1 min	Rhodamine 6G	80.5%	(Annadurai et al. 2003)

MW-assisted chemical activation exhibited enhanced activation efficiency. Acid-based activation was carried out with acetic acid, whereas oxidants like zinc chloride ($ZnCl_2$) and hydrogen peroxide (H_2O_2) were employed in the MW-assisted chemical activation process. The activation of waste activated sludge, papaya leaf stem, epicarp of *Ricinus communis* (RC), biochar, etc., showed an efficient and effective activation of carbonaceous material by MW heating.

The adsorption capacity of activated carbon from papaya leaf stem prepared by microwave-induced chemical activation with $ZnCl_2$ was investigated for the removal of malachite green and reactive orange dyes from aqueous solution (Mohan Raj and Asaithambi 2014). The adsorption capacity was 352 mg of malachite green (MG) and/or 163 mg reactive orange (RO)/g of activated carbon. In another study, MW-assisted activation in presence of $ZnCl_2$ adsorption capacity epicarp of *R. communis* (RC), an agricultural waste material, for the removal of MG was investigated using MW-assisted activation RC (MRC) and MW-assisted $ZnCl_2$ activated RC (ZRC). The adsorption capacity showed twofold increase for the removal of MG by ZRC compared to that of MRC (Makeswari and Santhi 2013). Tremendous improvement in the adsorption capacity was reported in the activation of biochar, the by-product obtained from biomass pyrolysis, by microwave in presence of H_2O_2. H_2O_2 impregnation followed by microwave treatment of biochar was investigated for the removal of Remazol brilliant blue (RBB) and methylene blue (MB) (Nair and Vinu 2016). The activated biochar produced at optimum condition, i.e., H_2O_2 impregnation time of 24 h and microwave power of 600 W (HC), exhibited nanostructure with narrow and deep pores and high specific surface area. The maximum adsorption capacity of RBB by HC was 1.5 times higher than that of non-activated biochar. On the other hand, there was twofold increase in MB adsorption capacity when HC was used as adsorbent instead of non-activated biochar. The maximum capacity of untreated biochar for RBB and MB was 63 and 42 mg/g, respectively, whereas the adsorption capacity with HC was obtained as 91 and 83 mg/g, respectively, for RBB and MB.

The activated carbon obtained from agricultural waste product by MW heating and chemical activation also showed good adsorption capacity for the removal of dye from wastewater (Foo and Hameed 2012a; b). Waste activated sludge, formed from conventional treatment plant, is a good adsorbent for heavy metals and various organics. However, the use of original sludge suffers the disadvantages of continuous change in quality and the odor emission during storage. On the other hand, the MW thermal treatment of sludge with a MW output power of 800 W in presence of acetic acid (10% w/w) showed improved adsorption efficiency of the synthetic dye, rhodamine 6G, from wastewater (Annadurai et al. 2003). Experimental results indicated higher adsorption efficiency at higher pH and temperature levels.

5.4.2 MW-Based Regeneration of Activated Carbon (AC)

Owing to the high porosity resulting in the increased surface area and surface activity, GAC and powdered activated carbon (PAC) effectively adsorb various classes of dyes (Gupta and Suhas 2009). However, the exhausted GAC/PAC needs to be regenerated for reuse and also to eliminate the sludge pollution. The thermal, chemical, and biological methods are commonly adopted for the regeneration of pollutant-loaded GAC (Purkait et al. 2007).

Thermal regeneration of saturated carbon is achieved by heating to a very high temperature (800–1000 °C) in mildly oxidizing conditions (Cong and Wu 2007). Chemical regeneration is usually carried out by desorption of adsorbates using specific solvents or by decomposition of adsorbed species using oxidizing chemical agents under either subcritical or supercritical conditions (Ryu et al. 2000). The regeneration efficiency by this method depends mainly on the solubility of the adsorbed substances and the effect of pressure on the chemical structure of carbon (Foo and Hameed 2009). Instead, the biological regeneration of exhausted carbon relies on the degradation of adsorbed pollutants by the disperse culture of microorganism (Salvador and Jimenez 1996). This can be achieved by mixing the saturated carbon with proper bacterial culture for sufficiently long contact time (Roy et al. 1999). However, the process is slow, sensitive to stringent operating conditions, and can be applied only to biodegradable pollutants adsorbed on the carbonaceous material. The use of these processes is challenging due to the high-energy consumption, long regeneration time, and reduction in the adsorption capacity. On the other hand, MW-assisted GAC regeneration possesses the advantages of short reaction time, energy savings, and improvement in the adsorption capacity of GAC.

Microwave heating restored the adsorption capacity of GAC and PAC effectively in several cycles of dye wastewater treatment (Table 5.2). MW regenerated GAC, obtained by immersing the spent GAC in water and subsequent MW

Table 5.2 MW-assisted regeneration of carbon for dye removal from wastewater

S. No.	Carbon compound	Regeneration method	Dye	Removal at optimum condition	Reference
1	GAC	Spent GAC was immersed it in water and subsequently exposed to microwave irradiation at a power of 385 W for 2 min	Methyl orange (MO)	~74.1% even after 6 cycles of usage	(Xu et al. 2014)
2	GAC	Reactive black 5-intermediate loaded GAC was regenerated by at MW power of 800 W, GAC/water of 20 g/L, and irradiation time of 30 s	Reactive Black 5	Effective regeneration of dyes for 6 cycles of operation	(Chang et al. 2010)
3	GAC	Dye-exhausted GAC was regenerated with MW irradiation for 5 min at 850 W	Acid Orange 7	Improved adsorption rate compared to that of virgin GAC due to pore-size distribution and surface modification	(Quan et al. 2004)
4	Activated carbon loaded with zinc acetate (AC-Zn)	MW heating of AC-Zn to a temperature of 850 °C for 35 min using nitrogen flow as shielding gas, followed by regeneration using CO_2 gas flow with temperature of 800 °C and duration of 30 min	Congo red (CR)	The regenerated AC-Zn revealed an adsorption of 16.72 mg of CR/g of regenerated adsorbent	(Qu et al. 2016)

irradiation at optimum conditions, showed about 74% removal of methyl orange (MO) from wastewater even after six cycles of usage (Xu et al. 2014). Similar result was obtained for Reactive Black 5 (RB5) and Acid Orange 7 (AO7) removal from wastewater using regenerated GAC. The adsorption of RB5 intermediates of electrocoagulation process on GAC indicated a tremendous decrease in the chemical oxygen demand (COD) of the wastewater (Chang et al. 2010). Subsequently, the GAC loaded with RB5 intermediates was regenerated by MW irradiation at a MW output power of 800 W and GAC/water ratio of 20 g/L. The adsorption capacity of GAC could be regenerated with 30 s of MW irradiation. Moreover, the adsorption capacity could be effectively regenerated for six cycles of operation. Alternatively, AO7 adsorption capacity tested for four cycles using regenerated GAC indicated a higher adsorption rate than that of virgin GAC due to pore-size distribution and surface modification (Quan et al. 2004).

On the other hand, MW regeneration was also effective for the regeneration of metal-loaded GAC. AC loaded with zinc acetate (AC-Zn) resulted from the adsorption of zinc acetate from vinyl acetate synthesis industry was regenerated by MW using different activation agents, namely steam, CO_2, and steam–CO_2 (Qu et al. 2016). Thus, prepared AC–zinc oxide (AC-ZnO) absorbent was investigated for the adsorptive removal of Congo red from wastewater. The result indicated a maximum adsorption capacity of 16.72 mg/g of AC-ZnO activated by CO_2 at optimum conditions.

5.5 Solvent-Phase Microwave Heating for Dye Wastewater Treatment

Dielectric loss heating is mainly responsible for the microwave heating of a solution. Solvent-phase MW heating of carbonaceous material resulted in the catalytic degradation of dyes from wastewater.

Solvent-phase MW heating aims in MW-induced catalytic technology by absorbing MW radiation by the strong MW absorbent (carbonaceous material) and transferring the energy to the pollutants causing the cleavage of complex bonds and thus the degradation. When activated carbon is used as MW-absorbing material, hot spots are formed on the surface which causes selective heating, spur in the molecular rotation, and finally decreases the activation energy. However, degradation of dye from wastewater by solvent-phase MW heating is still in the stage of infancy (Table 5.3).

Microwave irradiation with activated carbon was highly effective for the degradation of Congo red in aqueous solution. An efficiency of 97.88% was obtained for Congo red removal from wastewater by MW-PAC treatment under optimum conditions, i.e., 2.0 g/L PAC and 2.5-min microwave irradiation. A comparable degradation efficiency of Congo red (96.5%) was obtained within 1.5 min by increasing the dosage of PAC to 3.6 g/L (Zhang et al. 2007).

Treatment of Argazol blue containing wastewater with ferrous-modified GAC in presence of MW with an output power of 500 W for 5 min resulted in 98.3% decoloration and 96.8% COD removal from the dye wastewater (Park et al. 2010).

Table 5.3 MW-assisted degradation of dye from wastewater

S. No.	Carbon compound	Dye	Removal at optimum condition (%)	Reference
1	PAC	Congo red	97.88	(Zhang et al. 2007)
2	GAC	Argazol blue	98.3	(Guang 2004)

The decoloration and mineralization are caused by the synergetic effect of adsorption of dye from wastewater to GAC and subsequent degradation by MW irradiation.

5.6 Conclusion

MW irradiation possesses several advantages such as rapid and volumetric heating, increased reaction rates, reduced activation energies, and high energy efficiency. MW irradiation was used in dye wastewater treatment in two ways: (1) for solid-phase heating of activation or regeneration of carbonaceous material used as adsorbent in the treatment of dye wastewater and (2) solvent-phase heating for the degradation of adsorbed pollutant from the carbon surface. The dipolar polarization is mainly responsible for the activation of carbonaceous material or regeneration of spent carbon in solid-phase heating. Instead, dielectric loss heating is mainly responsible for solvent-phase heating by MW. MW-assisted activation in presence of activation agents like acid and oxidant improved the surface structure and porosity of carbonaceous material. On the other hand, the MW-based regeneration of spent carbon gave effective treatment efficiency of dye wastewater treatment for several cycles. Solvent-phase MW heating for adsorption and subsequent degradation of dye was seldom employed for the dye wastewater treatment. However, the method gave rapid and high treatment efficiency due to the synergetic effect of adsorption on carbon and thermal degradation by MW. The carbonaceous materials in different forms such as carbon supported on catalyst, carbon nanotubes, and carbon nanofilms can be coupled with MW to improve the treatment efficiency as well as the separability of the material in the solvent-phase degradation process.

References

Annadurai G, Juang RS, Yen PS, Lee DJ (2003) Use of thermally treated waste biological sludge as dye absorbent. Adv Environ Res 7:739–744

Chang S-H, Wang K-S, Liang H-H, Chen H-Y, Li H-C, Peng T-H, Su Y-C, Chang C-Y (2010) @@Treatment of reactive black 5 by combined electrocoagulation–granular activated carbon adsorption–microwave regeneration process. J Hazard Mater 175:850–857

Cong YQ, Wu ZC (2007) Electrocatalytic generation of radical intermediates over lead dioxide electrode doped with fluoride. Chin Chem Lett 18(8):1013–1016

Eskicioglu C, Terzian N, Kennedy KJ, Droste RL, Hamoda M (2007) Athermal effect of microwave for enhancing digestibility of waste activated sludge. Water Res 41(11):2457–2466

Foo KY, Hameed BH (2009) Recent developments in the preparation and regeneration of activated carbons by microwaves. Adv Colloid Interface Sci 149:19–27

Foo KY, Hameed BH (2012a) Adsorption characteristics of industrial solid waste derived activated carbon prepared by microwave heating for methylene blue. Fuel Process Technol 99:103–109

Foo KY, Hameed BH (2012b) Mesoporous activated carbon from wood sawdust by K_2CO_3 activation using microwave heating. Biores Technol 111:425–432
Gupta VK, Suhas (2009) Application of low-cost adsorbents for dye removal—a review. J Environ Manage 90:2313–2342
Hidaka H, Saitou A, Honjou H, Hosoda K, Moriya M, Serpone N (2007) Microwave-assisted dechlorination of polychlorobenzenes by hypophosphite anions in aqueous alkaline media in the presence of Pd-loaded active carbon. J Hazard Mater 148:22–28
Horikoshi S, Serpone N (2011) Microwave frequency effect(s) in organic chemistry. Mini-Rev Org Chem 8:299–305
Jones DA, Lelyveld TP, Mavrofidis SD, Kingman SW, Miles NJ (2002) Microwave heating applications in environmental engineering—a review. Resour Conserv Recycl 34:75–90
Kim T, Lee J, Lee K-H (2014) Microwave heating of carbon-based solid materials. Carbon Lett 15 (1):15–24
Lee CK, Low S, Chow CW (1990) Chrome sludge as an adsorbent for color removal. Environ Technol 17:1023–1028
Makeswari M, Santhi T (2013) Removal of malachite green dye from aqueous solutions onto microwave assisted zinc chloride chemical activated epicarp of *Ricinus communis*. J Water Resour Prot 5:222–238
Mall ID, Upaphyay SN (1995) Removal of basic dye from wastewater boiler bottom ash. Indian J Environ Health 37:1–4
Mohan Raj R, Asaithambi M (2014) Microwave assisted chemical activation of papaya leaf stem activated carbon for the removal of dyes from aqueous solution. J Environ Nanotechnol 3:91–95
Nair V, Vinu R (2016) Peroxide-assisted microwave activation of pyrolysis char for adsorption of dyes from wastewater. Biores Technol 216:511–519
National Research Council (NRC) (1994) Microwave processing of materials. National Academy Press, New York
Park SH, Kim S-J, Seo S-G, Jung S-C (2010) Assessment of microwave/UV/O_3 in the photo-catalytic degradation of bromothymol blue in aqueous nano TiO_2 particles dispersions. Nanoscale Res Lett 5:1627–1632
Peng W, Guo-yu Z, Si-peng J, Hui-jun M, Guang, H (2004) Microwave induced oxidation process for dye wastewater treatment. In: Proceedings of IWA world water congress, Marrakech, Morocco
Purkait MK, Maiti A, Das Gupta S, De S (2007) Removal of congo red using activated carbon and its regeneration. J Hazard Mater 145:287–295
Qu W, Hu Q, Zhu Y, Peng J, Zhang L (2016) Microwave-assisted regeneration of spent activated carbon containing zinc acetate and its application for removal of congo red. Desalin Water Treat 57(58):28496–28511
Quan X, Liu X, Bo L, Chen S, Zhao Y, Cui X (2004) Regeneration of acid orange 7-exhausted granular activated carbons with microwave irradiation. Water Res 38:4484–4490
Remya N, Lin J-G (2011) Current status of microwave application in wastewater treatment—a review. Chem Eng J 166(3):797–813
Remya N, Lin J-G (2015) Microwave-granular activate carbon (MW-GAC) system for carbofuran degradation: analysis of characteristics and recyclability of spend GAC. Desalin Water Treat 53:1621–1631
Robinson T, McMullan G, Marchant R, Nigam P (2001) Remediation of dyes in textile effluent: a critical review on current treatment technologies with a proposed alternative. Biores Technol 77:247–255
Roy D, Maillacheruvu K, Mouthon J (1999) Bioregeneration of granular activated carbon loaded with 2,4-D. J Environ Sci Health Part B 34:769–791
Ryu YK, Kim KL, Lee CH (2000) Adsorption and desorption of n-hexane, methyl ethyl ketone, and toluene on an activated carbon fiber from supercritical carbon dioxide. Ind Eng Chem Res 39(7):2510–2518

Salvador F, Jimenez CS (1996) A new method for regenerating activated carbon by thermal desorption with liquid water under subcritical conditions. Carbon 34(4):511–516

Xu D, Cheng F, Zhang Y (2014) Degradation of methyl orange in aqueous solution by microwave irradiation in the presence of granular-activated carbon. Water Air Soil Pollut 225:1983

Zhang Z, Shan Y, Wang J, Ling H, Zang S, Gao W, Zhao Z, Zhang H (2007) Investigation on the rapid degradation of congo red catalyzed by activated carbon powder under microwave irradiation. J Hazard Mater 147(1–2):325–333

Part IV
Various Techniques of Water Remediation

Chapter 6
Treatment Technologies for Emerging Organic Contaminants Removal from Wastewater

Sunita J. Varjani and M. Chaithanya Sudha

Abstract Of late, the existence of a group of contaminants identified as emerging contaminants (ECs) is of continued concern for health and safety of humans and ecosystem. Among various emerging contaminants, the emerging organic contaminants (EOCs) are components that form a significant part of products used by humans, viz. pharmaceuticals and personal care products (PPCP), pesticides, veterinary medicines, fire retardants and other industrial products. Endocrine-disrupting chemicals (EDCs) are also often portrayed in a category of their own. The significant number of individual EOCs is present in wastewater treatment plant (WWTP) due to domestic and industrial effluent emissions. Estimation of EOCs gains more importance due to complexity involved in analysing them. Physicochemical and biological properties are vital components deciding the fate of EOCs in wastewater treatment and upon their discharge into nature. Therefore, it is very essential to collect scientific data on the fate and effects of EOCs to assess the environmental risk. The current chapter analyses adverse effects of EOCs to water consumers and discusses various potential removal processes. The use of physical, chemical and biological methods such as adsorption, ultrafiltration, electrodialysis, Fenton reaction, photo catalysis, biofilter and removal in algal wastewater treatment ponds, and their efficiencies in removal of EOCs are reviewed.

Keywords Emerging organic contaminants · Wastewater · Treatment technologies

S. J. Varjani
Gujarat Pollution Control Board, Sector-10A, Gandhinagar 382010, Gujarat, India

M. Chaithanya Sudha (✉)
VIT University, Vellore, Tamil Nadu, India
e-mail: mchaithanyasudha@gmail.com

S. Bhattacharya et al. (eds.), *Water Remediation*, Energy, Environment, and Sustainability, https://doi.org/10.1007/978-981-10-7551-3_6

6.1 Introduction

Until the origin of the nineties, non-polar hazardous compounds, i.e. persistent organic pollutants (POPs) and heavy metal, were a focus of interest and awareness as priority pollutants. Today, these compounds are less relevant for the industrialized countries, since a dramatic reduction of emissions has been achieved through the adoption of appropriate measures and therefore, elimination of the dominant sources of pollution. Organic compounds previously not known to be significant in freshwater, in terms of distribution and/or concentration, are now being more widely detected as analytical techniques improve (Farre et al. 2012). These compounds, which have the potential drop to cause known or suspected adverse ecological or human health effects, are often collectively referred to as ECs. The United States Geological Survey (USGS) defines an emerging contaminant as "any synthetic or naturally occurring chemical or any micro-organism that is not commonly monitored in the environment but has the potential to enter the environment and cause known or suspected adverse ecological and (or) human health effects. In some cases, release of emerging chemical or microbial contaminants to the environment has likely occurred for a long time, but may not have been recognized until new detection methods were developed. In other cases, synthesis of new chemicals or changes in use and disposal of existing chemicals can create new sources of ECs".

ECs arise from many types of products and various chemicals such as sucralose, artificial sweeteners, nanomaterials, perfluorinated compounds, drinking water and swimming pool disinfection by-products, sunscreens and UV filters, flame retardants, benzotriazoles and benzothiazoles, siloxanes, naphthenic acids, musks, algal pollutants, and ionic liquids and prions (Zgheib et al. 2011). With advancements inside the chemical industry, the type of compounds being released to the surroundings which can be doubtlessly dangerous to human beings and ecosystem over the long time period is expected to grow substantially over time (Cassani and Gramatica 2015).

ECs are not new chemicals; they might be substances that have been available on earth for quite a while, however, whose presence and significance are perceived as of late. EOCs are principally synthetic organic chemicals that have been recently distinguished in natural habitats. ECs are a large and relatively new group of unregulated compounds that might be derived from domestic, trade and industrial sources. It can conceivably cause injurious issues in amphibians and human life (Fig. 6.1). They are mostly identified and detected in metropolitan sewage, day-by-day household items, pharmaceutical production plants, wastewater, hospitals, landfills and natural aquatic environment (Fig. 6.1).

The largest number of ECs is man-made chemicals, which include new classes of halogenated persistent organic pollutants, pharmaceuticals, chemicals in personal care products and natural steroid hormones. ECs are differentiated from conventional persistent organic pollutants like polychlorinated biphenyls and organochlorine pesticides (e.g. DDT) via their bioactive properties. EOCs are components or active ingredients in products that are frequently used by humans.

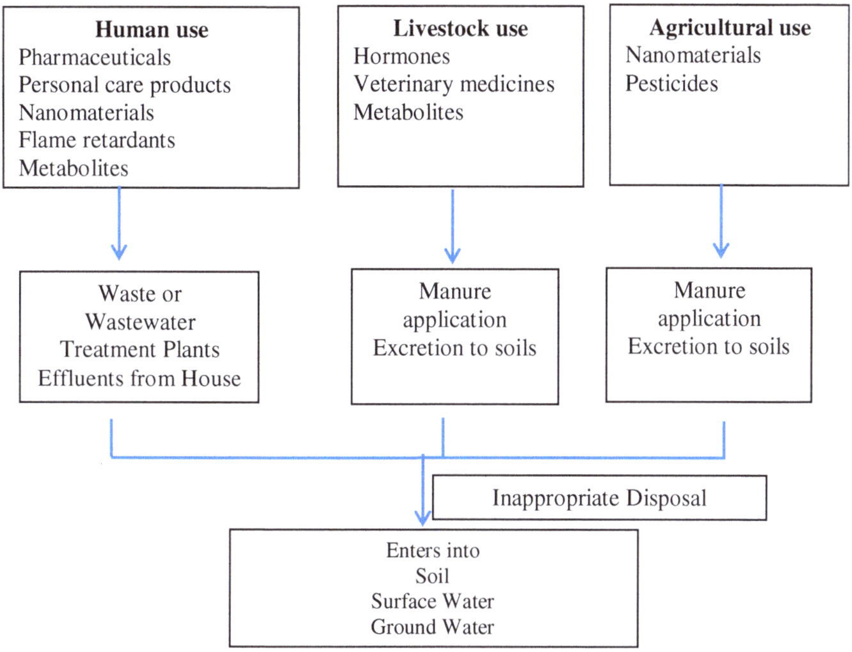

Fig. 6.1 Routes of entry of ECs to the environment

The predominant groups of EOCs include PPCP, veterinary medicines, fire retardants and other industrial products, and new era pesticides. In comparison to industrial chemicals, EOCs are sourced from products that are utilized in relatively small amounts. However, due to the fact they are used by many people on multiple occasions every day, the total amounts released to the environment can be significant.

Therefore, the estimation of EOCs gains more importance due to the complexity involved in analysing them. Physicochemical and biological properties are vital components deciding the fate of EOCs amid wastewater treatment and upon their discharge into nature. Therefore, it is very essential to collect scientific data on the fate and effects of EOCs to assess the environmental risk. The current paper analyses adverse effects of EOCs to water consumers and discusses the various potential removal processes. The use of physical, chemical and biological methods such as adsorption, ultrafiltration, electrodialysis, Fenton reaction, photo catalysis, biofilter and removal in algal wastewater treatment ponds, and their efficiencies in removal of EOCs are reviewed.

6.2 Emerging Contaminants in Environment

ECs will be released into the aquatic environments (surface and groundwater), agricultural environment, soil environment and food materials via a number of routes. The ECs from soil may be transported to water bodies by leaching, run-off and drainage processes. The extent of the transport is dependent on the persistence of the EC and on how it interacts with soil and sediment particles. These contaminants can also be released to the aquatic environment from other sources such as emissions from WWTPs and industrial effluents (Pal et al. 2010). The ECs from the emissions of human medicines and human personal care products are likely to be more significant than from agricultural emissions.

6.2.1 Pharmaceuticals

Pharmaceuticals are a critical group of ECs, and their presence in drinking water has generated significant issues regarding the risk of oestrogenic and other adverse impacts on humans and fauna (Liu et al. 2015). Approximately 3000 distinctive substances are estimated to be used as pharmaceutical ingredients, including painkillers, antibiotics, antidiabetics, beta blockers, contraceptives, lipid regulators, antidepressants and impotence pills. Only a small subset of these ECs has been investigated in environmental research. The large-scale use of prescribed drugs has also improved their presence in surface water, groundwater, wastewater and storm water run-off in urban regions (McGrane 2016; Campanha et al. 2015).

Pharmaceuticals find their way to the environment through various pathways (Fig. 6.2); however, the major way is through the discharge of raw and treated sewage from residential users or medical centres (Rahman et al. 2009). The excretion through urine and faeces and extensively metabolized drugs are released into the receiving waters. On the other side, the topically applied pharmaceuticals (when washed off) and the expired and unused ones (when disposed directly to trash or sewage), as well as released from drug manufacturing plants (Larsson et al. 2007; Li et al. 2008), may pose a direct risk to the environment due to the fact that they enter sewage in their unmetabolized and effective form (Ruhoya and Daughton 2008). Besides the wastewater effluents discharged to water bodies or reused for irrigation, biosolids used in agriculture as soil amendment or disposed to landfill are another significant route of pharmaceuticals to the environment and drinking water resources (Barron et al. 2008).

The Indian pharmaceutical industry positions third in terms of volume production in the world with approximately 250 large-scale units and more than 8,000 small-scale units in current operation including central public sector holdings (CCI 2015). Major pharmaceutical production centres in India are located in Ahmedabad, Bangalore, Hyderabad and Mumbai and supply these products to Europe, the USA and other regions of the world (Larsson 2008). A total of 19 pharmaceuticals have

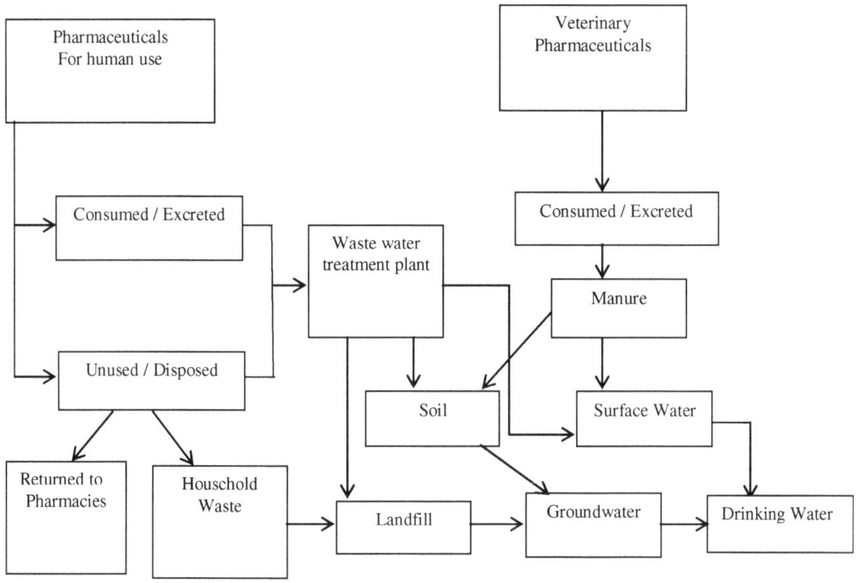

Fig. 6.2 Main routes of pharmaceuticals to the environment

been measured so far in surface water, wastewater and drinking water sources all over India. The monitoring has been done so far in Kaveri, Vellar and Tamiraparani Rivers in southern India, wastewater treatment plants in New Delhi and Ujjain in central India, and surface waters (lakes, river) and the wastewater of Hyderabad in north-western India.

6.2.2 *Antibiotics, Biocides and Pesticides*

The important concern in relation to antibiotics, biocides and pesticides is the development of bacterial resistance after their release into the environment (Zhang et al. 2015; Petit et al. 2015) and the unfavourable effect on the biodegradation of plant materials, which disrupts the primary food chain in aquatic ecosystems (Richardson and Ternes 2014). The term "pesticide" refers to chemical substances used for agricultural purposes, whereas the term "biocide" refers to chemical compounds used in urban environments (Margot et al. 2015). Biocides are specifically used in bituminous roof sealing membranes and external facades or for grass management and weed control. During rain events, biocides and pesticides are incorporated in surface and groundwater via storm water run-off (Weston et al. 2015; Tang et al. 2015; Ricardo Bento and Bentes 2015).

Endosulfan, hexachloro-cyclohexane (HCHs) and DDT are the most consumed pesticides in India, among which the latter two categories together account for 67%

of total pesticide consumption (Kumari et al. 2001). Others include phorate, chlorpyriphos, methyl parathion and phosphamidon (Bhushan et al. 2013). HCHs and DDTs are included in the list of priority substances reported in Directive 2008/105/EC (European Commission 2008) and their maximum environmental quality standards (EQSs) in inland surface water bodies are 0.2 and 2.5 ng/L, respectively (European Commission 2013).

6.2.3 *Personal Care Products*

These chemical substances are found extensively in urban environments and encompass fragrances, sunscreens, insect repellents and antifungal agents (Pal et al. 2014). Since these compounds are designed for external usage, no metabolic changes occur in their chemical structure and are easily released into aquatic environments. Their presence in urban run-off and groundwater has also increased notably in recent years (Weston et al. 2015; Tang et al. 2015; Ricardo Bento and Bentes 2015). Triclosan (TCS), a disinfectant, has been widely used for almost 30 years in soaps, deodorants, skin creams and toothpastes (McAvoy et al. 2002). TCS is also used in some commercial products such as footwear (hosiery and soles of shoes), acne creams, toys and kitchen utensils. It is reported to be in the top 10 most commonly detected organic pollutants in wastewater (Kolpin et al. 2004; Halden and Paull 2005). Triclosan and parabens (methyl, ethyl, propyl and butyl) have been so far investigated in the Kaveri, Vellar and Tamiraparani Rivers as well as at the Pichavaram mangrove of southern India (Ramaswamy et al. 2011). The compounds were also monitored in sediments of these rivers.

6.2.4 *Phthalates and Plastic*

Phthalates and phthalate acid ester (PAE) are frequently used in the manufacturing of PVC products, furnishings, textiles, toys, food packaging and medicinal products (Mackintosh et al. 2004). Human beings may be exposed to phthalates through inhalation, packed food items, drinking water and use of personal care products, especially those having oral contact. Therefore, more attention is given to estimation of concentration of phthalates in air, food items and drinking water sources. Currently there are approximately 26,000 registered plastic processing units in India, and the demand for plastic raw material increased more than twofold from 3.3 to 6.8 million tons in the first decade of the twenty-first century (i.e. between 2000–2001 and 2009–2010). A few studies are carried out that measured the concentration of phthalates in surface water (Gomti River, Kaveri River), drinking water (tap and bottle water in southern part of Delhi, India) and wastewater in India (15 sewage treatment plants in northern India) (Das et al. 2014; Selvaraj et al. 2015; Srivastava et al. 2010; Gani et al. 2016). Das et al. (2014) monitored the concentration of 15 phthalates in bottled and purified drinking water from Okhla, India, and the Jawaharlal Nehru University (JNU) campus in southern Delhi, India.

6.3 Emerging Contaminants—Effects on Human Health and Ecosystem

The ECs such as PPCPs, pesticides and hormones have detrimental effects on human and wildlife endocrine systems. Natural attenuation and conventional treatment methods are not capable of removing these micropollutants from wastewater and surface and drinking water and are reported to bioaccumulate in macroinvertebrates, other organisms in the aquatic food web and humans (Ruhí et al. 2015). The critical hazards of these compounds on the human health and ecosystem are given in Table 6.1.

Table 6.1 Effects of ECs on aquatic and terrestrial organisms

Emerging contaminants	Reported effect(s)
Anorexic treatments	Hormonal effect in crabs and crayfish
Oestrogens	Endocrine-disrupting effects on fish
Parasiticides	Effects on insect development and physiology; effect on rate of dung decomposition
Antibacterials	Selection of antibacterial resistance; impacts on the structure of soil microbial communities; inhibition of growth blue-green algae and aquatic plants; biochemical effects in fish
Anti-inflammatory	Stimulation of growth of blue-green algae and inhibition of growth of aquatic plants; effects on structure of fish hepatocytes
Lipid regulator	Biochemical effects in fish
Analgesic	Biochemical effects in fish
Beta blocker	Biochemical effects in fish
Lipid regulator	Biochemical effects in fish
Antianxiety drug	Effects on development of invertebrates
Cardiac glycoside	Effects on development of invertebrates
Calcium channel blocker	Effects on development of invertebrates
Carbamazepine and diazepam	Growth is inhibited for both Daphnia magna and midges
Beta blockers	Chronic toxic effects have been found in fish
Hormones	Male infertility
Butylated hydroxy toluene (BHT) and butylated hydro-xyanisole (BHA)	Carcinogenic effects

(continued)

Table 6.1 (continued)

Emerging contaminants	Reported effect(s)
Sunscreens/UV filters	Estrogenic effects; potential for developmental toxicity
Flame retardants	Developmental toxicity, carcinogenicity and hormone disruption
Bisphenol A (BPA)	Harmful effects on infants and young children
Alkylphenols and ethoxylates	Negative effect on aquatic life by acting as endocrine disruptors
N-nitrosodimethylamine (NDMA)	Probable human carcinogen
Benzotriazoles and benzothiazoles	In vitro oestrogenic effects
Perfluorinated compounds	Environmental consequences and potential human health effect

Source Boxall (2012) and Birkholz et al. (2014)

6.4 Analytical Methodologies for ECs Quantification

One of the significant constraints in the analysis of emerging contaminants remains the lack of methods for quantification of low concentrations. The prerequisite for proper risk assessment and monitoring of the quality of waste, surface and drinking waters is the accessibility of multi-residual techniques that allow measurement at the low ng/l level (or even below that). However, these compounds have obtained little interest because they are not on regulatory lists as environmental pollutants. However, today analytical methodology for different groups of emerging contaminants is being developed, and an increasing number of techniques are reported in the literature (Petrovic et al. 2003a, b).

In the last two decades, several analytical methods have been developed for the determination of different classes of ECs in environmental samples (water, sediment, soil and biota). Generally, the identification and quantification of ECs at low concentrations in complex environmental matrices requires analytical methods of high sensitivity and selectivity, which usually depend on liquid or gas chromatography (LC or GC) coupled to mass spectrometry (MS). The application of advanced low- or high-resolution MS instruments in environmental analysis has allowed the determination of a broader range of compounds and, thus, a more comprehensive assessment of environmental contaminants.

The preferred analytical method is based on target analysis of preselected compounds of interest, using tandem MS instruments. Over the years, a slow shift from class-specific methods to multi-residue methods for simultaneous analysis of a large number of target compounds, belonging to different classes, has took place. For instance, Robles-Molina et al. (2014) developed a multi-residue method based on solid-phase extraction and liquid chromatography-time-of-flight-mass spectrometry for the simultaneous determination of over 400 priority and emerging pollutants in water and wastewater. On the other hand, Pintado-Herrera et al. (2014)

used atmospheric pressure gas chromatography-time-of-flight-mass spectrometry (APGC-ToF-MS) for the simultaneous determination of 102 regulated and emerging contaminants in aqueous samples after stir bar sorptive extraction (SBSE).

A number of multi-residue methods have been developed for the analysis of multiclass pharmaceuticals, illicit drugs, pesticides, PFC. Gros et al. (2012) developed a method to determine 81 pharmaceutical residues, covering various therapeutic groups, and some of their main metabolites, in surface and treated waters (influent and effluent wastewaters, river, reservoirs, sea and drinking water); Baker and Kasprzyk-Hordern (2011) defined a multi-residue method for the environmental monitoring of 65 stimulants, opioid and morphine derivatives, benzodiazepines, antidepressants, dissociative anaesthetics, drug precursors, human urine indicators, and their metabolites in wastewater and surface water. Masiá et al. (2013) developed a method for the examination of 43 polar pesticides in environmental waters.

The current diagnostic approach has another disadvantage. The majority of analytical techniques focus mainly on parent target compounds and rarely comprise metabolites and transformation products, which on occasion can be extra toxic and persistent than the original compounds. One cause for that trend is that the majority of transformation products are not known and many of those that are known are not commercially available or are too expensive. However, each one of the relevant contaminants, metabolites and transformation products that may be encountered in the aquatic environment are difficult to be incorporated in any targeted multi-residue method, making therefore a strong case for the application of non-target screening protocols using high-resolution mass spectrometry (HR-MS) (Chiaia-Hernandez et al. 2013; Eichhorn et al. 2012).

In contrast to triple quadrupole mass spectrometers, which function at unit resolution for precise target analytes, HR-MS units which include time of flight (TOF), quadrupole time of flight (QqTOF) and Orbitrap mass spectrometers are able to obtain full-scan mass spectra at high decision for all analytes, therefore, the unique potential of retrospective data analysis for compounds not included in the first data processing. Accurate mass complete-scan MS allows analysis of a wide variety of compounds and their degradation products that fall outside the scope of target techniques. However, general screening for unknown materials is time-consuming and highly priced and is often shattered by problems, which includes lack of mass spectral libraries. Therefore, the fundamental challenge is to prioritize contaminants and determine on the significance of the chemical information. Effect-related analysis, focused on relevant compounds, in recent times appears to be the maximum appropriate manner to assess and study environmental contamination (Petrovic et al. 2016).

6.5 Treatment Technologies for Emerging Contaminants

The number of chemical groups constituting ECs is large and continues to grow as new chemicals are identified to be part of this classification. Many emerging trace organic contaminants, such as endocrine-disrupting chemicals, personal care products and pharmaceuticals are detected in wastewater, surface water, groundwater and drinking water. Attention to EOCs is escalating due to their potential undesirable health effects on humans and ecosystems (Jones et al. 2004; Bound and Voulvoulis 2004; Tran et al. 2013). Several EOCs are released into aquatic environments, mostly by effluents from municipal sewage treatment plants (Heberer 2002; Bound and Voulvoulis 2004) because they are not completely removed by conventional treatment technologies (Carballa et al. 2004; Díaz-Cruz and Barceló 2008; Reemtsma et al. 2010; Radke et al. 2009; Kasprzyk-Hordern et al. 2008).

Approximately 3000 different substances are estimated to be used as pharmaceutical ingredients, including painkillers, antibiotics, antidiabetics, beta blockers, contraceptives, lipid regulators, antidepressants and impotence drugs. Only a small subset of these ECs has been investigated in environmental studies. The large-scale use of pharmaceuticals has also increased their presence in surface water, groundwater, waste water and storm water run-off in urban areas. The chemicals from personal care products are found widely in urban environments and include fragrances, sunscreens, insect repellents and antifungal agents. Since these compounds are designed for external usage, no metabolic changes occur in their chemical structure and are easily released into aquatic environments. Their presence in urban run-off and groundwater has also increased significantly in recent years. Non-conventional water treatment technologies have changed over time as a result of new techniques being developed. These treatments can be broadly divided into physicochemical, biological treatment and advanced treatment processes (Table 6.2).

6.5.1 *Physicochemical Treatment*

The important physical and chemical methods used for the removal of emerging contaminants are discussed in this section. Physicochemical treatment as a coagulation–flocculation process was generally found to be unable to remove EDCs and PPCPs. However, coagulation combined with PAC adsorption has been reported as a very attractive option due to several advantages, such as an easy use and low capital investments (Dixon et al. 2011). The use of aluminium sulphate and ferric chloride as coagulants combined with UF has provided notably improvements in the permeate flux, with considerably reduction at the same time of turbidity and chemical oxygen demand (COD) (Wuo et al. 2004; Kim et al. 2005; Fan et al. 2008).

Table 6.2 Overview of physicochemical and biological treatment methods for ECs

Physicochemical methods	Biological methods
Chlorination	Activated sludge process
Sorption	Biological activated carbon
Adsorption	Biosorption
Coagulation	Biological trickling filter and biofilm reactor
Immobilization	Biological nitrification and denitrification
Photolysis	Membrane bioreactor
Photo catalysis	Constructed wetlands (CWs)
Photo degradation	Subsurface flow CWs
Fenton process	Surface flow (SF) CWs
Electro-Fenton processes	Hybrid CW systems
Photo-Fenton process	Waste stabilization ponds (WSPs)
Sonochemical reaction	High-rate algal ponds (HRAPs)
Electrodialysis	Fungal reactors
Ultrafiltration	Daphnia reactors
	Microalgae-/fungi-based treatment
	Phytoremediation

6.5.1.1 Chlorination

The less reactive species such as chlorine (gaseous chlorine and hypochlorite) and bromine have been used in wastewater treatment. The effect of chlorine on the removal of some of the ECs has been carried out by Noutsopoulos et al. (2014) using 1000 ng L^{-1} of each ECs pollutant after exposing initial chlorine dose of 11 mg L^{-1} for 60 min. The maximum removal efficiencies were 95 and 100%, respectively, for naproxen and diclofenac. The removal of EE2 by chlorination was found to be up to 100% within 10 min (Belgiorno et al. 2007). Moreover, it was observed that the reaction rate of chlorination process was three orders of magnitude lower than that of ozonation process during the removal of ECs such as amitriptyline hydrochloride, methyl salicylate and 2-phenoxyethanol (Real et al. 2014). In addition, chlorine and chlorine dioxide are potent oxidants which may produce some subproducts during wastewater treatment, and the degree of mineralization achieved is not acceptable (Rivera-Utrilla et al. 2013).

6.5.1.2 Ultrafiltration

Separation processes by membranes, such as ultrafiltration (UF), are technologies increasingly used in the field of water and wastewater treatments, and produce clear water suited for different applications. However, single UF is sometimes ineffective for the removal of most of these emerging contaminants due to the limited retention capacity of UF membranes as well as to the membrane fouling. Then, membrane filtration processes must be combined with pretreatments or post-treatments in order to produce a permeate that can be reused, such as coagulation and adsorption stages, which permit additional removals of organic compounds that play an important role in fouling phenomena. Specifically, powdered activated carbon (PAC) and granular activated carbon (GAC) in combination with UF are promising technologies for water.

The combinations of PAC adsorption and/or coagulation pretreatments followed by UF, as well as by an UF treatment followed by GAC adsorption post-treatment were investigated to remove the EOCs (Acero et al. 2017). The PAC pretreatment decreased membrane fouling, with the advantage that PAC was separated from the final effluent in the UF step. Low PAC dose in the range 10–50 mg L^{-1} in the adsorption pretreatment was enough in order to remove most of the ECs and to partially improve water quality parameters. The removal of 11 selected emerging contaminants (ECs, acetaminophen, metoprolol, caffeine, antipyrine, sulfamethoxazole, flumequine, ketorolac, atrazine, isoproturon, 2-hydroxybiphenyl and diclofenac) by micellar-enhanced ultrafiltration (MEUF) has been investigated by Acero et al. (2017). Anionic sodium dodecylsulphate (SDS), non-ionic surfactants Triton X-100 (TX-100) and Tween 20 (TW-20), and cationic surfactants cetylpyridinium chloride (CPC) and cetyl trimethyl ammonium bromide (CTAB) were used. The retention coefficients of the selected compounds were determined in order to evaluate the separation efficiency of ECs from surfactant micelles. It was found that cationic surfactants were more appropriate for the removal of negatively charged and hydrophobic ECs.

6.5.1.3 Immobilization

Immobilization methods are developed in order to facilitate enzyme recuperation and reusability and to increase enzymatic stability. Enzymes had been immobilized through numerous methods portrayed in several comprehensive reviews (Sheldon 2007; Zhou and Hartmann 2012). Basically, three methods can be distinguished, i.e. support or carrier binding, encapsulation or entrapment, and cross-linking (Durán et al. 2002; Sheldon 2007; Ba et al. 2013). Each technique possesses specific advantages and disadvantages relying on the targeted application. Carrier binding methods such as physical adsorption and covalent or ionic binding immobilize the enzymes on a support material. Adsorption happens by means of electrostatic interactions, van der Waals forces and even hydrogen bonding between enzymes and carrier surface. The immobilization of laccases for the treatment of wastewaters has been recently reviewed by Ba et al. (2013).

The removal of rising organic contaminants from municipal wastewater poses a significant challenge unsatisfactorily addressed by present wastewater treatment processes. Enzyme-catalysed transformation of EOC has been planned as a potential resolution to the present major environmental issue. Especially, laccases gained interest during this context in recent years owing to their broad substrate range and since they solely want molecular oxygen as a co-substrate. Artificial recharge improves many water quality parameters, but has only minor effects on recalcitrant pollutants. To advance the removal of those pollutants, added a reactive barrier at the bottom of an infiltration basin (Valhondo et al. 2014). This barrier contained aquifer sand, vegetable compost and clay and was coated with iron oxide dust.

The purpose of the compost was to sorb neutral compounds and release dissolved organic carbon. The release of dissolved organic carbon ought to generate a broad range of redox conditions to promote the transformation of emerging trace organic contaminants (EOCs). Iron oxides and clay increase the range of sorption site types. Valhondo et al. (2014) studied the effectiveness of the barrier by analysing the fate of nine EOCs. Water quality was monitored before and after constructing the reactive barrier. Installation of the reactive barrier led to nitrate-, iron- and manganese-reducing conditions in the unsaturated zone below the basin and within the first few metres of the saturated zone. Thus, the behaviour of most EOCs changed after installing the reactive barrier. The reactive barrier enhanced the removal of some EOCs, either markedly (sulfamethoxazole, caffeine, benzoylecgonine) or slightly (trimethoprim) and decreased the removal rates of compounds that are easily degradable under aerobic conditions (ibuprofen, paracetamol). The barrier had no remarkable effect on 1H-benzotriazole and tolyltriazole.

6.5.1.4 Photo Catalysis

Prieto et al. (2012) studied about the treatment of emerging contaminants in WWTP effluents by solar photo catalysis using low TiO_2 concentrations. A mild photocatalytic tertiary treatment for the removal of 52 ECs from WWTP biological secondary treatment effluents was demonstrated under realistic conditions (real wastewater with contaminants in the μg L^{-1} range). Although the low concentration of the catalyst limits the reaction rate due to the weak absorption of photons in current reactor designs, a higher extent of ECs degradation (>85%) is possible under natural sunlight when CPC reactors of larger diameters are used. The use of low concentrations of TiO_2 could be a good, economical and simple alternative for the tertiary treatment of WWTP and for the removal of persistent ECs which are not degraded by traditional WWTPs. A strategy for increasing the absorption of solar radiation involves the use of tubular reactors and CPC with a wider tube diameter to allow optimal absorption of photons.

6.5.1.5 Advanced Oxidation Process

Chemical oxidation methods such as ozonation/H_2O_2, UV photolysis/H_2O_2 and photo-Fenton processes were reported as the best processes for the removal of pesticides, beta blockers and pharmaceuticals. Ozonation and UV photo catalysis processes are highly effective in the removal of EDCs. The Fenton process has been observed to be the least effective among all types of conventional and AOPs treatment technologies. The removal of surfactants and PCPs has not yet been well studied by chemical processes. Finally, hybrid systems such as MBR followed by reverse osmosis, nanofiltration or ultrafiltration are better for the removal of EDCs and pharmaceuticals, but less effective in the removal of pesticides.

Advanced oxidation process (AOPs) may be the best choice when complete mineralization of organic component is required. Sonochemical reactions have been successfully employed for the degradation of various pollutants.

Methylparaben (MPB) which is the most utilized from paraben family is selected as a model emerging pollutant. The influence of common anions and some selected emerging contaminants that may coexist in typical water matrix on the degradation pattern is analysed alongside. Among the anions, carbonate presents a negative influence which is attributed to the competition for OH radical. Some ECs also showed negative impact on degradation as was clear from HPLC data. The intermediates, analysed by LC–Q-TOF–MS, include hydroxylated and hydrolytic products. Three major steps (aromatic hydroxylation, hydroxylation at the ester chain and hydrolysis) are proposed to involve in the reaction of OH radical with MPB which ultimately leads to mineralization. The major driving mechanisms of the degradation of MPB were found to be hydroxylation and hydrolysis. A 100% mineralization as indicated by the mass data clearly demonstrates the efficient degradation of this compound by sonochemical technique compared to any other AOP techniques.

6.5.1.6 Sonochemical Reaction

Sonochemical reactions have been efficaciously utilized for the degradation of different pollutants (Darbre and Harvey 2008; Byford et al. 2002). It is yet another route which makes use of OH radical for the degradation. In sonolysis, complete mineralization of organic pollutants is expected within a short span of time. However, this degradation protocol has not been endeavoured so far for the possible mineralization of parabens. In sonolysis, there are a few parameters which depend on the ultrasonic frequency that controls the degradation efficiency. The most vital being the number and size of bubbles, and the availability of OH in the interface and bulk liquid region (Yasui 2011). Finding an optimum frequency is critical where all these parameters correlate. The increase in applied frequency improves the accessibility of OH, however in the meantime brings about a diminishment of the full-size bubble (Petrier et al. 1992).

Subha et al. examined the sonolytic degradation of methylparaben (MPB) which is the most used of paraben family as an emerging pollutant. The impact of common anions and some selected emerging contaminants that may exit together in typical water matrix on the degradation pattern is analysed alongside. Among the anions, carbonate offers a poor affect which is ascribed to the opposition of OH radical. Three most important steps (aromatic hydroxylation, hydroxylation at the ester chain and hydrolysis) are proposed to involve in the reaction of OH radical with MPB which ultimately results in mineralization. The intensity of formation and decay of mono- and dihydroxy products of MPB in the presence of additives have also been evaluated. COD examination shows a rate decrease of 98% at 90 min of sonolysis and further increase in the degradation time resulted complete mineralization, which have become obtrusive from the mass spectrometric data. MTT test revealed significant decrease in the potential cytotoxicity.

6.5.1.7 Electrodialysis

The electrodialysis process uses a driving force of direct current power to transfer ionic species from the source water through cation and anion exchange membranes to a concentrate waste stream, creating a more diluted product water stream. Electrodialysis reversal (EDR) is a variation on the electrodialysis process, which uses periodic electrode polarity reversal to automatically clean membrane surfaces (Valero and Arbos 2010; Valero et al. 2013). Gabarróna et al. evaluated the occurrence and removal of pharmaceutically active compounds (PhACs), EDCs and related compounds in a drinking water treatment plant (DWTP) treating raw water from the Mediterranean Llobregat River. The DWTP combined conventional treatment steps with the world's largest electrodialysis reversal (EDR) facility. A total of 49 different PhACs, EDCs and related compounds were found above their limit of quantification in the influent of the DWTP, summing up to a total concentration of ECs between 1600 and 4200 ng/L. The oxidation using chlorine dioxide and GAC filters were found to be most efficient technologies for EC removal. However, despite the low concentration detected in the influent of the EDR process, it was also possible to demonstrate that this process partially removed ionized compounds, thereby constituting an additional barrier against EC pollution in the product. In the product of the EDR system, only 18 out of 49 compounds were quantifiable in at least one of the four experimental campaigns, showing in all cases removals higher than 65% and often beyond 90% for the overall DWTP process.

6.5.2 Biological Treatment Process

Biological wastewater treatment systems are considered as sustainable, cost-effective alternative to conventional wastewater treatment system(s). Biological systems can be used for treatment of urban sewage from small communities; however, it can also be used effectively to remove EOCs. EOCs are reported as a new group of unregulated contaminants which include PPCPs, some pesticides, veterinary products and industrial compounds among others that are thought to have long-term adverse effects on human health and ecosystems. The fate of EOCs in these systems is highly dependent on their physicochemical properties, such as their hydrophobicity (log Kow) and ionization state (pKa). Usually, aerobic pathways are more efficient for the removal of EOCs; nevertheless, some compounds can also be efficiently removed under anaerobic conditions. The presence of vegetation is a remarkable factor since in addition to the plant uptake; it enhances biodegradation mechanisms, as they release plant exudates and increase the amount of micro-organisms present in water. Environmental factors such as temperature and sunlight radiation should be considered, as they are limiting factors for microbe's activity and are related to plant photosynthesis.

6.5.2.1 Membrane Bioreactors

Biological treatment technologies were broadly carried out for the removal of ECs predominantly by using the mechanism of biodegradation. Biodegradation is the method by which large molecular weight ECs are degraded by means of micro-organisms such as bacteria, algal and fungi into small molecules (Garcia-Rodríguez et al. 2014), and even biomineralized to simple inorganic molecules consisting of water and carbon dioxide. In conventional biodegradation technique, micro-organisms use organic compounds as principal substrates for their cell growth and persuade enzymes for their assimilation (Tran et al. 2013). A few ECs are toxic and resistant against microbial growth hence inhibiting biodegradation, in which a growth substrate is wanted to sustain microbial growth for biodegradation, a procedure known as co-metabolism (Tran et al. 2013). Biodegradation methods have traditionally been utilized in wastewater treatment systems for the elimination of ECs. They can be divided into aerobic and anaerobic processes. Aerobic applications include activated sludge, membrane bioreactor and sequence batch reactor. Anaerobic methods include anaerobic sludge reactors and anaerobic film reactors.

Membrane bioreactor (MBR) is capable to effectively remove a wide variety of ECs including compounds that can be resistant to activate sludge process and constructed wetland (Luo et al. 2014; Radjenovic et al. 2009). This may be accomplished because of sludge retention on the membrane surface which can promote microbial degradation and physical retention of all molecules larger than the molecular weight of the membrane. The elimination of ECs in MBR system can be influenced by sludge age, concentration and presence of anoxic and anaerobic compartments, composition of wastewater, operating temperature, pH and conductivity (Luo et al. 2014). Ozonation technique is most extensively used together with MBR process. Adsorption and biodegradation were found to be liable for the removal of ECs by MBR treatment.

In assessment to conventional activated sludge process, MBR can remove higher amount of EDCs from wastewater (Maeng et al. 2013). PCPs which include salicylic acid and propyl parabene were removed by around 100% in MBR system. The removal of pesticides, for example atrazine, dicamba, fenoprop, 2,4-D and pentachlorophenol from wastewater by MBR was not satisfactory except for triclosan removal which can be up to 99%. A few ECs such as beta blockers can be removed by this process at 70–80%, and atendol can be eliminated by up to 97%. On account of pharmaceuticals removal, MBR showed a mixed performance.

In general, the elimination of some slowly degradable pharmaceutical drugs which include antibiotics and analgesics in MBRs is better due to the relatively long sludge ages, which leads to the growth of distinct microbial communities in MBRs in comparison to activated sludge plants. But the removal of pharmaceuticals such as anti-inflammatory and stimulant drugs by MBR and activated sludge is comparable. For the removal of pharmaceuticals in MBR, their performance followed the order of analgesics > antibiotics > anti-inflammatory and stimulants > other pharmaceuticals. The overall trend of ECs removal by MBR can be composed as EDCs > PCPs > beta blockers > pharmaceuticals > pesticides (Ahmed et al. 2017).

6.5.2.2 Activated Sludge Process

Different biological processes are reported to enhance the removal efficiency of ECs. For example, conventional activated sludge process has been reported to show better removal efficiency for surfactants, EDCs and PCPs than trickling filter and biofilm reactors, nitrification and denitrification processes. Biological activated carbon process has been reported with enhanced efficiencies in the removal of pesticides, analgesics and antibiotics. MBR process has been reported as highly effective technique for the removal of EDCs, PCPs and beta blockers than constructed wetland. Novel microalgae-based technology has shown the highest efficiency to remove many categories of ECs especially pharmaceutical sand PCPs, although no data were reported on their removal of beta blockers, antibiotics and surfactants.

Ozonation followed by biological activated carbon hybrid system has been reported effective in removal of pesticides, beta blockers and pharmaceuticals. Ozonation followed by ultrasound hybrid system has been reported to remove up to 100% of some pharmaceuticals such as salicylic acid, ibuprofen, naproxen, acetaminophen, cocaethylene, benzoylecgonine, enalapril, norben zoylecgonine, ketoprofen, atorvastatin, bezafibrate, clindamycin, sulfamethazine and 4-aminoantipyrine. Other hybrid systems based on activated sludge followed by ultrafiltration or activated sludge followed by gamma radiation are cost-effective to remove certain EDCs, pesticides and analgesic pharmaceuticals. Hybrid systems using ultrafiltration, activated carbon followed by ultrasound process can be a better option to remove a wide range of ECs but may not be cost-effective.

6.5.2.3 Aerobic and Anaerobic Degradation

The biodegradation studies with EOCs, such as PPCPs, have mainly focused on aerobic conditions. One of the reasons probably is the assumption that the aerobic degradation is more energetically favourable than degradation under strictly anaerobic conditions. Certain aerobically recalcitrant contaminants, however, are biodegraded under strictly anaerobic conditions, and little is known about the organisms and enzymatic processes involved in their degradation. The anaerobic biotransformation reactions for a variety of well-studied, structurally simple mono- or oligo functional contaminants (SMOCs) bearing one or a few different functional groups/structural moieties were well reviewed.

Treatment of primary wastewater effluent containing a mixture of nine pharmaceuticals model compounds (acetaminophen ACM, antipyrine ANT, caffeine CAF, ketorolac KET, metoprolol MET, sulfamethoxazole SFX, carbamazepine CARB, hydrochlorothiazide HCT and diclofenac DIC) has been carried out by means of aerobic biodegradation followed by different advanced oxidation processes (solar heterogeneous photocatalysis with TiO_2, solar photo-Fenton and ozonation). Only ACM and CAF were completely removed from municipal wastewater in aerobic biological experiments.

Though anaerobic degradation succeed in several environmental compartments of waste water treatment, biodegradation studies with EOCs, which include pharmaceuticals and personal care products; have specially focused on aerobic situations for their effective removal from the water and wastewater. One of the motives probable is the assumption that the aerobic degradation is extra energetically beneficial than degradation underneath strictly anaerobic situations. Certain aerobically recalcitrant contaminants, however, are biodegraded beneath strictly anaerobic conditions and little is understood approximately the organisms and enzymatic methods worried of their degradation. Ghattas et al. (2017) reviewed a comprehensive survey of characteristic anaerobic biotransformation reactions for a variety of well-studied, structurally rather simple contaminants (SMOCs) bearing one or some different functional group moieties.

Furthermore, it summarizes anaerobic degradation studies of greater complex contaminants with several functional groups (CMCs), in soil, sediment and wastewater treatment. The anaerobic conditions are able to promote the transformation of several aerobically persistent contaminants. The most prevalent reactions observed were reductive dehalogenations and the cleavage of ether bonds. Thus, it turns clear that the transferability of deterioration mechanisms deduced from culture studies of SMOCs are expecting the degradation of CMCs, consisting of EOCs, in environmental matrices is hampered due the more complex chemical structure bearing one-of-a-kind useful corporations, distinct environmental conditions (e.g. matrix, redox, pH), the microbial community (e.g. version, competition) and the low concentrations common for EOCs.

6.5.2.4 Algal Ponds

The sorption of ECs to the solids found and generated during wastewater treatment is highly dependent on (a) EC structure, viz. hydrophobicity and functional groups available for chemisorption (b) properties of solid(s) acting as sorbent, viz. hydrophobicity and functional groups available for chemisorption and (c) environmental conditions. Recent research has significantly progressed knowledge about removal of ECs in algal wastewater treatment (WWT) ponds. However, due to the vast array of ECs detected in WWT and complexity of ecology and environmental conditions occurring in algal WWT ponds, this area of research is still in its infancy. The long HRTs in algal ponds may allow removal mechanisms with slow kinetics to become significant. This may, for example, allow time for hydrolysis, or biodegradation following de-conjugation. The biodegradation of ECs by algae has mainly been reported in laboratory studies and is based on monocultures of algae grown in specific media. This potential must therefore be verified in case of wastewater. Very few studies have quantified the potential of EC sorption to algal biomass.

In 2013, photodegradation of ECs in algal ponds has been reported, which is expected due to the large surface-area-to-volume ratios of algal ponds. However, few studies have properly isolated the significance of photodegradation from other removal mechanisms in algal ponds. Due to the presence of organic compounds absorbing and scattering of light, the indirect photodegradation dominated over direct photolysis for most of the ECs in algal ponds, and all photodegradation processes are likely to occur between 10 and 20 cm from the surface. Light-based mechanisms may also reduce the risks associated with antibiotic resistance in algal WWT through disinfection processes and the destruction of antibiotic resistance genes. The large daily and seasonable fluctuations in temperature, pH and DO commonly experienced in algal ponds should impact rates and quantitative significance of hydrolysis, sorption, biodegradation and photodegradation.

6.5.2.5 Biological Trickling Filter and Biofilm Reactor

A biological trickling clear out is a three-segment system with constant biofilm providers. Wastewater enters the bioreactor through a distribution area, trickles downward over the biofilm surface, and air moves upward or downward in the 0.33 section (Daigger and Boltz 2011). Biotrickling filters had been used in WWTPs for many years in the removal of biochemical oxygen demand (BOD), chemical oxygen demand (COD), pathogen decontamination, smell and air pollutant control; however, their application to ECs removal has no longer come to be wide practice (McLamore et al. 2008; Naz et al. 2015; Wang et al. 2006). Some biostrategies such as activated sludge, aerated lagoon and trickling filters have stated very extraordinary removal efficiencies from nearly entire elimination to no removal of a few prescribed drugs from different wastewater sources (Lin et al. 2009).

Kasprzyk-Hordern et al. (2009) monitored 55 pharmaceuticals, PCPs, EDCs and illicit drugs in the course of wastewater treatment through trickling filter out and activated sludge techniques from South Wales, UK, over a length of five months. They concluded that the activated sludge remedy was a far greater efficient system than trickling clear out beds for the elimination of natural micropollutants. Overall, out of 55 prescribed drugs and PCPs studied, only some have been characterized via low removal performance (<50%) at some point of activated sludge remedy. In comparison, the WWTP using trickling filter out beds resulted in, on common, less than 70% elimination of all 55 PPCPs studied with half of them not being eliminated, while the WWTP using activated sludge remedy gave a better higher removal performance of over 85% (Kasprzyk-Hordern et al. 2009; Svenson et al. 2003). Hence, there may be a want to increase or modify the existing biotrickling method to acquire better and steadier removal efficiency for a huge variety of ECs.

A pilot-scale biofilter was set up with natural manganese oxides as carrier materials at a local STP to treat the real secondary effluent (Zhang et al. 2017). This innovative and simple reactor was on-site operated for approximately 500 days. Some EOCs were effectively removed after adaptation, including 10,11-dihydro-10,11-dihydroxycarbamazepine (98%), gabapentin (97%), tramadol (93%), carbamazepine (91%), benzotriazole (91%), sulfamethoxazole (88%), erythromycin (86%). By contrast, the removal of others can be obtained without adaptation, including diclofenac (91%), carboxy-acyclovir (91%), iomeprol (89%), 1-hydroxybenzotriazol (87%), 40-hydroxydiclofenac (86%), acyclovir (73%), tolyltriazole (70%). Overall, 80% of the total mass of 15 detected EOCs was eliminated from the secondary effluent. In addition, 53% of UV254 was removed from wastewater, indicating the aromatic content was damaged to a certain extent.

6.6 Conclusion

The EOCs are components that form a significant part of products used by humans, such as PPCPs, veterinary medicines, fire retardants and other industrial products, and pesticides. EDCs are also often portrayed in a category of their own. The significant number of individual EOCs is present in WWTP effluent. Estimation of EOCs gains more importance due to the complexity involved in analysing them. Physicochemical and biological properties are vital components deciding the fate of EOCs amid wastewater treatment and upon their discharge into nature. Chemical oxidation methods such as ozonation/H_2O_2, UV photolysis/H_2O_2 and photo-Fenton processes have been found to be the best processes for the removal of pesticides, beta blockers and pharmaceuticals. EDCs were better removed by MBR, activated sludge and aeration processes among different biological processes. Surfactants, EDCs and personal care products (PCPs) can be well removed by activated sludge process.

References

Acero JL, Benitez FJ, Real FJ, Teva F (2017) Removal of emerging contaminants from secondary effluents by micellar-enhanced ultrafiltration. Sep Pur Technol 181:123–131

Ahmed MB, Zhou JL, Ngo HH, Guo W, Thomaidis NS, Jiang Xu (2017) Progress in the biological and chemical treatment technologies foremerging contaminant removal from wastewater: A critical review. J Hazard Mater 323:274–298

Ba S, Arsenault A, Hassani T, Jones JP, Cabana H (2013) Laccase immobilization and insolubilization: from fundamentals to applications for the elimination of emerging contaminants in wastewater treatment. Crit Rev Biotechnol 33:404–418

Baker DR, Kasprzyk-Hordern B (2011) Multi-residue analysis of drugs of abuse in wastewater and surface water by solid-phase extraction and liquid chromatography-positive electrospray ionisation tandem mass spectrometry. J Chromatogr A 1218(21):1620–1631

Barron L, Tobin J, Paull B (2008) Multi-residue determination of pharmaceuticals in sludge and sludge enriched soils using pressurized liquid extraction, solid phase extraction and liquid chromatography with tandem mass spectrometry. J Environ Monit 10:353–361

Belgiorno V, Rizzo L, Fatta D, Della Rocca C, Lofrano G, Nikolaou A, Naddeo V, Meric S (2007) Review on endocrine disrupting-emerging compounds inurban wastewater: occurrence and removal by photocatalysis and ultrasonicirradiation for wastewater reuse. Desalination 215 (1–3):166–176

Bound J, Voulvoulis N (2004) Pharmaceuticals in the aquatic environment—a comparison of risk assessment strategies. Chemosphere

Boxall ABA (2012) New and emerging water pollutants arising from agriculture, OECD study, water quality and agriculture: meeting the policy challenge. Environment Department, University of York, United Kingdom

Byford JR, Shaw LE, Drew MGB, Pope GS, Sauer MJ, Darbre PD (2002) Oestrogenic activity of parabens in MCF7 human breast cancer cells. J Steroid Biochem Mol Biol 80(2002):49–60

Campanha MB, Awan AT, de Sousa DNR, Grosseli GM, Mozeto AA, Fadini PS (2015) A 3-year study on occurrence of emerging contaminants in an urban stream of São Paulo State of Southeast Brazil. Environ Sci Pollut Res 22(2015):7936–7947

Carballa M, Omil F, Lema JM, Llompart M, Garc´ia-Jares C, Rodŕiguez I, Gómez M, Ternes T (2004) Behavior of pharmaceuticals, cosmetics and hormones in a sewage treatment plant. Water Res 38(12):2918–2926. https://doi.org/10.1016/j.watres.2004.03.029

Cassani S, Gramatica P (2015) Identification of potential PBT behavior of personal care products by structural approaches. Sustain Chem Pharm 1:19–27

CCI (Corporate Catalyst India) (2015) A brief report on pharmaceutical industry in India

Chiaia-Hernandez AC, Krauss M, Hollender J (2013) Screening of lake sediments for emerging contaminants by liquid chromatography atmospheric pressure photoionization and electrospray ionization coupled to high resolution mass spectrometry. Environ Sci Technol 47(2):976–986

Daigger GT, Boltz JP (2011) Trickling filter and trickling filter-suspended growth process design and operation: a state-of-the-art review. Water Environ Res 83:388–404

Darbre PD, Harvey PW (2008) Paraben esters: review of recent studies ofendocrine toxicity, absorption, esterase and human exposure, and discussionof potential human health risks. J Appl Toxicol 28:561–578

Das MT, Ghosh P, Thakur IS (2014) Intake estimates of phthalate esters for South Delhi population based on exposure media assessment. Environ Pollut 189:118–125

Díaz-Cruz MS, Barceló D (2008) Trace organic chemicals contamination in ground water recharge. Chemosphere 72(3):333–342. https://doi.org/10.1016/j.chemosphere.2008.02.031

Dixon MB, Richard Y, Ho L, Chow CWK, O'Neill BK, Newcombe G (2011) A coagulation-powered activated carbon-ultrafiltration—multiple barrier approach for removing toxins from two Australian cyanobacterial blooms. J Hazard Mater 186(2011):1553–1559

Durán N, Rosa MA, D'Annibale A, Gianfreda L (2002) Applications of laccases and tyrosinases (phenoloxidases) immobilized on different supports: a review. Enzym Microb Technol 31:907–931

Eichhorn P, Pérez S, Barceló D (2012) Time-of-flight mass spectrometry versus orbitrap-based mass spectrometry for the screening and identification of drugs and metabolites: is there a winner? In: Rodriguez Fernandez-Alba A (ed) TOF-MS within food and environmental analysis, vol 58. Comprehensive analytical chemistry. Elsevier, New York, pp 217–272

European Commission (2008) Directive 2008/105/EC of the European parliament and of the council of 16 December 2008 on environmental quality standards in the field of water policy, amending and subsequently repealing Council Directives 82/176/EEC, 83/513/EEC, 84/156/EEC, 84/491/EEC, 86/280/EEC and amending Directive 2000/60/EC of the European Parliament and of the Council

European Commission (2013) Directive 2013/39/EU of the European parliament and of the council of 12 August 2013 amending Directives 2000/60/EC and 2008/105/EC as regards priority substances in the field of water policy

Fan L, Nguyen T, Roddick FA, Harris JL (2008) Low-pressure membrane filtration of secondary effluent in water reuse: pretreatment for fouling reduction. J Membr Sci 320(2008):135–142

Farre M, Kantiani L, Petrovic M, Perez S, Barcelo D (2012) Achievements and future trends in the analysis of emerging organic contaminants in environmental samples by mass spectrometry and bioanalytical techniques. J Chromatogr

Gaffney VJ et al (2014) Analysis of pharmaceutical compounds in waters by SPE-UPLC-ESI-MS/MS Quím. Nova 37(1):138e149

Gani KM, Rajpal A, Kazmi AA (2016) Contamination level of four priority phthalates in North Indian wastewater treatment plants and their fate in sequencing batch reactor systems. Environ Sci Process Impacts 18(3):406–416

Garcia-Rodríguez A, Matamoros V, Fontàs C, Salvadó V (2014) The ability of biologically based wastewater treatment systems to remove emerging organic contaminants—a review. Environ Sci Pollut Res 21(2014):11708–11728

Ghattas AK, Fischer F, Wick A, Ternes TA (2017) Anaerobic biodegradation of (emerging) organic contaminants in the aquatic environment. Water Res 116:268–295

Gros M, Rodríguez-Mozaz S, Barceló D (2012) Fast and comprehensive multi-residue analysis of a broad range of human and veterinary pharmaceuticals and some of their metabolites in surface and treated waters by ultra-high-performance liquid chromatography coupled to quadrupole-linear ion trap tandem mass spectrometry. J Chromatogr A 1248:104–121

Halden RU, Paull DH (2005) Co-occurrence of triclocarban and triclosan in US water resources. Environ Sci Technol 39(6):1420–1426

Heberer T (2002) Occurrence, fate, and removal of pharmaceutical residues in the aquatic environment: a review of recent research data. Toxicol Lett 131(1–2):5–17. https://doi.org/10.1016/S03784274(02)00041-3

Jones OAH, Voulvoulis N, Lester JN (2004) Potential ecological and human health risks associated with the presence of pharmaceutically active compounds in the aquatic environment. Crit Rev Toxicol 34(4):335–350. https://doi.org/10.1080/10408440490464697

Kasprzyk-Hordern B, Dinsdale RM, Guwy AJ (2008) The occurrence of pharmaceuticals, personal care products, endocrine disruptors and illicit drugs in surface water in south wales, UK. Water Res 42(13):3498–3518. https://doi.org/10.1016/j.watres.2008.04.026

Kasprzyk-Hordern B, Dinsdale RM, Guwy AJ (2009) The removal of pharmaceuticals personal care products, endocrine disruptors and illicit drugs during wastewater treatment and its impact on the quality of receiving waters. Water Res 43(2009):363–380

Kim SH, Moon SY, Yoon CH, Yim SK, Cho JW (2005) Role of coagulation in membrane filtration of wastewater for reuse. Desalination 173(2005):301–307

Kolpin DW, Skopec M, Meyer MT, Furlong ET, Zaugg SD (2004) Urban contamination of pharmaceuticals and other organic wastewater contaminants to streams during differing flow conditions. Sci Total Environ 328(1–3):119–130

Kumari B, Kumar R, Kathpal TS (2001) An improved multiresidue procedure for determination of pesticides in vegetables. Pestic

Larsson DGJ (2008) Drug production facilities: an overlooked discharge source for pharmaceuticals to the environment. In: Kummerer K (ed) Pharmaceuticals in the environment: sources, fate, effects and risks. Springer, New York, pp 37–42

Larsson DGJ, de Pedro C, Paxeus N (2007) Effluent from drug manufactures contains extremely high levels of pharmaceuticals. J Hazard Mater 148:751–755

Leung H et al (2013) Pharmaceuticals in tap water: human health risk assessment and proposed monitoring framework in China. Environ Health Perspect 121(7):839e846

Li D, Yang M, Hu J, Zhang Y, Chang H, Jin F (2008) Determination of penicillin G and its degradation products in a penicillin production wastewater treatment plant and the receiving river. Water Res 42:307–317

Lin AYC, Yu TH, Lateef SK (2009) Removal of pharmaceuticals in secondary wastewater treatment processes in Taiwan. J Hazard Mater 167(2009):1163–1169

Liu A, Goonetilleke A, Egodawatta P (2015) Role of rainfall and catchment characteristics on urban stormwater quality. Springer, Singapore. http://dx.doi.org/10.1007/978-981-287-459-7

Luo Y, Guo W, Ngo HH, Nghiem LD, Hai FI, Zhang J, Liang S, Wang XC (2014) A review on the occurrence of micropollutants in the aquatic environmentand their fate and removal during wastewater treatment. Sci Total Environ 473(2014):619–641

Mackintosh CE, Javier AM, Hongwu J, Hoover N, Chong A, Ikonomou MG, Gobas Frank APC (2004) Distribution of Phthalate Esters in a Marine Aquatic Food Web: Comparison to Polychlorinated Biphenyls. Environ Sci Technol 38(7):2011–2020. http://pubs.acs.org/author/Mackintosh%2C+Cheryl+E

Maeng SK, Choi BG, Lee KT, Song KG (2013) Influences of solid retention time, nitrification and microbial activity on the attenuation of pharmaceuticalsand estrogens in membrane bioreactors. Water Res 47:3151–3162

Margot J, Rossi L, Barry DA, Holliger C (2015) A review of the fate of micropollutants in wastewater treatment plants. Wiley Interdiscip Rev Water 2(2015):457–487

Masiá A, Ibáñez M, Blasco C, Sancho JV, Picó Y, Hernández FF (2013) Combined use of liquid chromatography triple quadrupole mass spectrometry and liquid chromatography quadrupole time-of-flight mass spectrometry in systematic screening of pesticides and other contaminants in water samples. Anal Chim Acta 761:117–127

McAvoy DC, Schatowitz B, Jacob M, Hauk A, Eckhoff WS (2002) Measurement of triclosan in wastewater treatment systems. Environ Toxicol Chem 21(7):1323–1329

McGrane SJ (2016) Impacts of urbanisation on hydrological and water quality dynamics, and urban water management: a review. Hydrol Sci J 61:2295–2311

McLamore E, Sharvelle S, Huang Z, Banks K (2008) Simultaneous treatment of gray water and waste gas in a biological trickling filter. Water Environ Res 80(2008):2096–2103

Naz I, Saroj DP, Mumtaz S, Ali N, Ahmed S (2015) Assessment of biological trickling filter systems with various packing materials for improved wastewater treatment. Environ Technol 36(2015):424–434

Noutsopoulos C, Mamais D, Mpouras T, Kokkinidou D, Samaras V, Antoniou K, Gioldasi M (2014) The role of activated carbon and disinfection on the removal of endocrine disrupting chemicals and non-steroidal anti-inflammatory drugs from wastewater. Environ Technol 35 (2014):698–708

Pal A, Gin KYH, Lin AYC, Reinhard M (2010) Impacts of emerging organic contaminants on freshwater resources: review of recent occurrences, sources, fate and effects. Sci Total Environ 408:6062–6069

Pal A, He Y, Jekel M, Reinhard M, Gin KYH (2014) Emerging contaminants of public health significance as water quality indicator compounds in the urban water cycle. Environ Int 71 (2014):46–62. http://dx.doi.org/10.1016/j.envint.2014.05.025

Petit F, Denamur E, Clermont O, Leclercq R, Deloffre J, Cattoir V, Oberlé K, Budzinski K, Berthe T (2015) Fate of antibiotics and antibiotic-resistant fecal bacteria in water and sediments from the contamination source to the estuary: Impact and/or resilience? Resilience to contamination by antibiotics. In: Mar. Product. Perturbations Resil. Socio-Ecosystems, Springer International Publishing, Cham, pp 79–91. http://dx.doi.org/10.1007/978-3-319-13878-7_9

Petrier C, Jeunet A, Luche JL, Reverdy G (1992) Unexpected frequency effects on the rate of oxidative processes induced by ultrasound. J Am Chem Soc 114(1992):3148–3150

Petrovic M, Diaz A, Ventura F, Barcelo D (2003a) Occurrence and removal of estrogenic short chain ethoxy nonylphenolic compound and their halogenated derivatives during drinking water production. Env Sci Technol 27(2003):4442–4448

Petrović M, Gonzalez S, Barceló D (2003b) Analysis and removal of emerging contaminants in wastewater and drinking water. Trends Anal Chem 22(10):685–696

Petrovic M, Sabater S, Elosegi A, Barceló D (2016) Emerging contaminants in river ecosystems. Occurrence and effects under multiple stress conditions. https://doi.org/10.1007/978-3-319-29376-9

Pintado-Herrera MG, González-Mazo E, Lara-Martín PA (2014) Atmospheric pressure gas chromatography-time-of-flight-mass spectrometry (APGC-ToF-MS) for the determination of regulated and emerging contaminants in aqueous samples after stir bar sorptive extraction (SBSE). Anal Chim Acta 851:1–13

Prieto-Rodriguez L, Miralles-Cuevas S, Oller I, Agüera A, Puma GL, Malato S (2012) Treatment of emerging contaminants in wastewater treatment plants (WWTP) effluents by solar photocatalysis using low TiO2 concentrations, J Hazard Mater 211–212:131–137

Radjenovic J, Petrović M, Barceló D (2009) Fate and distribution of pharmaceuticals in wastewater and sewage sludge of the conventional activated sludge (CAS) and advanced membrane bioreactor (MBR) treatment. Water Res 43(2009):831–841

Radke M, Lauwigi C, Heinkele G, M¨urdter TE, Letzel M (2009) Fate of the antibiotic sulfamethoxazole and its two major human metabolites in a water sediment test. Environ Sci Technol 43(9):3135–3141. https://doi.org/10.1021/es900300u

Rahman MF, Yanful EK, Jasim SY (2009) Endocrine disrupting compounds (EDCs) and pharmaceuticals and personal care products (PPCPs) in the aquatic environment: implications for the drinking water industry and global environmental health. J Water Health 7:224–243

Ramaswamy BR, Shanmugam G, Velu G, Rengarajan B, Larsson DGJ (2011) GC-MS analysis and ecotoxicological risk assessment of triclosan, carbamazepine and parabens in Indian rivers. J Hazard Mater 186(2–3):1586–1593

Real FJ, Benitez JF, Acero JL, Casas F (2014) Comparison between chlorination and ozonation treatments for the elimination of the emerging contaminants amitriptyline hydrochloride, methyl salicylate and 2-phenoxyethanol in surface waters and secondary effluents. J Chem Technol Biotechnol 90:1400–1407

Reemtsma T, Miehe U, Duennbier U, Jekel M (2010) Polar pollutants in municipal wastewater and the water cycle: occurrence and removal of benzotriazoles. Water Res 44(2):596–604

Ricardo Bento CM, Bentes I (2015) Urban land-cover, urbanization type and implications for storm water quality: vila real as a case study. J Hydrogeol Hydrol Eng 4:2. https://doi.org/10.4172/2325-9647.1000122

Richardson SD, Ternes TA (2014) Water analysis: emerging contaminants and current issues. Anal Chem 86(6):2813–2848

Rivera-Utrilla J, Sánchez-Polo M, Ferro-García MÁ, Prados-Joya G, Ocampo-Pérez R (2013) Pharmaceuticals as emerging contaminants and their removal from water: a review. Chemosphere 93(2013):1268–1287

Robles-Molina J, Lara-Ortega FJ, Gilbert-López B, García-Reyes JF, Molina-Díaz A (2014) Multi-residue method for the determination of over 400 priority and emerging pollutants in water and wastewater by solid-phase extraction and liquid chromatography-time-of-flight mass spectrometry. J Chromatogr A 1350:30–43

Ruhí A, Acuña V, Barceló D, HuertaB, MorJR, Rodríguez-Mozaz S, Sabater S (2015) Bioaccumulation and trophic magnification of pharmaceuticals and endocrine disruptors in a Mediterranean river food web. Sci Total Environ 540(2015):250–259

Ruhoya ISR, Daughton CG (2008) Beyond the medicine cabinet: an analysis of where and why medications accumulate. Environ Int 34:1157–1169

Selvaraj KK, Sundaramoorthy G, Ravichandran PK, Girijan GK, Sampath S, Ramaswamy BR (2015) Phthalate esters in water and sediments of the Kaveri River, India: environmental levels and eco-toxicological evaluations. Environ Geochem Health 37(1):83–96

Sheldon RA (2007) Enzyme immobilization: the quest for optimum performance. Adv Synth Catal 349:1289–1307

Srivastava A, Sharma VP, Tripathi R, Kumar R, Patel DK, Mathur PK (2010) Occurrence of phthalic acid esters in Gomti River sediment, India. Environ Monit Assess 169(1–4):397–406

Svenson A, Allard AS, Ek M (2003) Removal of estrogenicity in Swedish municipal sewage treatment plants, Water Res 37(2003):4433–4443

Tang T, Boënne W, Desmet N, Seuntjens P, Bronders J, van Griensven A (2015) Quantification and characterization of glyphosate use and loss in a residential area. Sci Total Environ 517 (2015):207–214

Valero F, Arbos R (2010) Desalination of brackish river water using electrodialysis reversal (EDR). Control of the THMs formation in the Barcelona (NE Spain) area. Desalination 253:170–174

Valero F, Barceló A, Medina E, Arbós R (2013) Barcelona: three years of experience in brackish water desalination using EDR to improve quality. New O&M procedures to reduce low-value work and increase productivity, Desalin. Water Treat 51(2013):1137–1142

Valhondo C, Carrera J, Ayora C, Barbieri M, Nödler K, Licha T, Huerta M (2014) Behavior of nine selected emerging trace organic contaminants in an artificial recharge system supplemented with a reactive barrier, Environ Sci Poll Res 21(20):11832–11843

Vieno NM, Tuhkanen T, Kronberg L (2006) Removal of pharmaceuticals in drinking water treatment: effect of chemical coagulation. Env Technol 27(2006):183–192

Wang C, Li J, Wang B, Zhang G (2006) Development of an empirical model for domestic wastewater treatment by biological aerated filter. Process Biochem 41(2006):778–782

Weston DP, Schlenk D, Riar N, Lydy MJ, Brooks ML (2015) Effects of pyrethroid insecticides in urban runoff on Chinook salmon, steelhead trout, and their invertebrate prey. Environ Toxicol Chem 34:649–657

Wuo WS, Vigneswaran S, Ngo HH, Chapman H (2004) Experimental investigation of adsorption–flocculation–microfiltration hybrid system in wastewater reuse. J Membr Sci 242(2004):27–35

Yasui K (2011) Fundamentals of acoustic cavitation and sonochemistry. In: Ashok Kumar M (ed) Theoretical and experimental sonochemistry involving inorganic systems. Springer, Netherlands, pp 1–29

Zgheib S, Moilleron R, Saad M, Chebbo G (2011) Partition of pollution between dissolved and particulate phases: what about emerging substances in urban stormwater catchments? Water Res 45:913–925

Zhang QQ, Ying GG, Pan CG, Liu YS, Zhao JL (2015) Comprehensive evaluation of antibiotics emission and fate in the river basins of China: source analysis, multimedia modeling, and linkage to bacterial resistance. Environ Sci Technol 49(2015):6772–6782

Zhang Y, Zhu H, Szewzyk U, Lübbecke S, Geissen SU (2017) Removal of emerging organic contaminants with a pilot-scale biofilter packed with natural manganese oxides. Chem Eng J 317(1):454–460. http://www.sciencedirect.com/science/article/pii/S1385894717302589

Zhou Z, Hartmann M (2012) Recent progress in biocatalysis with enzymes immobilized on mesoporous hosts. Top Catal 55:1081–1100

Chapter 7
Role of Photo-catalysis in Water Remediation

Ankur Gupta, Kunal Mondal and Satyam Kumar

Abstract Photo-catalysis is one of the most explored phenomenons to obtain potable water in remote areas. The term "photo-catalysis" is understood as a chemical reaction which is caused by photo-absorption of a matter also known as "photo-catalyst", which remains unaltered throughout the reaction. In this process, water molecules adsorbed over the surface of photo-catalysts in normal environments are used to participate in the reaction to decompose pollutants in water. Since the photo-catalytic reactions are possible in air, the oxygen reduction is the significant phenomenon in photo-catalytic reduction. This chapter aims to cover the fundamental aspect of photo-catalytic phenomenon and its significant role in water purification. Explicit examples on the amalgamation of photo-catalysis and various photo-catalytic set-ups developed are also illustrated.

Keywords Photo-catalysis · Water purification · Nano-material
Batch reactor

7.1 Introduction

Water pollution is one of the major concerns of the world population. People in remote places are unable to accomplish the crucial requirement of clean drinking water. Among various harmful substances present in wastewater, some of them include organic substances. Industrial wastewater might furthermore include traces of heavy metals and organic substances. Researchers as well as common citizens based on their regional circumstances have attempted number of conventional

A. Gupta (✉) · S. Kumar
School of Mechanical Sciences, Indian Institute of Technology Bhubaneswar, Bhubaneswar, Odisha, India
e-mail: ankurgupta@iitbbs.ac.in

K. Mondal
Department of Chemical and Biomolecular Engineering, North Carolina State University, 911 Partners Way, Raleigh 27695, NC, USA

S. Bhattacharya et al. (eds.), *Water Remediation*, Energy, Environment, and Sustainability, https://doi.org/10.1007/978-981-10-7551-3_7

methods for water purification, viz. filtration through winnowing sieve, filtration through cloth, filtration through clay vessels, using some specific plants, using bleach, iodine to kill germs and many more. Out of these explored methods, photo-catalytic treatment of wastewater is also considered as important technique in water remediation. Here, in this article, we will focus to understand the role of photo-catalysis in water treatment and possible growth happened in this area. So firstly, let us understand the term "catalysis". Catalysis can be defined as a process which is used to change the rate of reaction, by accelerating or decelerating the reaction using catalyst. It takes part in the reaction to change the rate of reaction and returns in its original form without being consumed at the completion of the reaction. It can be used for many numbers of cycles without getting deteriorated.

Photo-catalysis is the process of altering the rate of photo-chemical reaction in the presence of light. It absorbs photons and generates electron–hole pair which further produces free radicals. It is a type of redox reaction which promotes both oxidation and reduction simultaneously. "Photo-synthesis", word which most of us come to know while understanding the natural phenomenon for plant's growth, is an example of photo-catalysis using chlorophyll as photo-catalyst. Water splitting process, also known as artificial synthesis, is used to break water into hydrogen and oxygen using photo-catalyst. It has been divided into two subgroups: homogeneous photo-catalysis (in which reactants and photo-catalysts both remain in the same phase) and heterogeneous photo-catalysis (in which reactant and photo-catalysts remain in different phase). Metals are generally not chosen for photo-catalyst due to the absence of the band gap and non-metals due to high band gap. Semiconductor material with reasonable band gap is suitable for photo-catalyst (Lee et al. 2016; Newton et al. 2001).

7.1.1 Characteristics of Photo-catalytic Material

Choice of photo-catalytic material should be based on the any material having reasonable band gap, so that electron can be excited with some suitable source of energy having wavelength of ultraviolet or visible range from electromagnetic spectrum. These wavelength requirements can easily be made available. These ranges are not very high as required in insulators or very low as in metals. It should easily transport the charge generated so that it can take part in the reaction to generate radicals. Further, structure or crystallinity of the material will affect both the above properties. It should have high surface area so that it can have more number of active sites. It must be stable so that it can be used many times.

When a photon strikes on the semiconducting photo-catalyst surface, electron absorbs the energy of the photon and excited to jump from valence band to conduction band generating an electron–hole pair in conduction and valence band, respectively. To carry out the photo-catalytic effect, the combination of electron–hole pair must be inhibit or the time of combination should be long enough to carry out the photo-catalytic reaction.

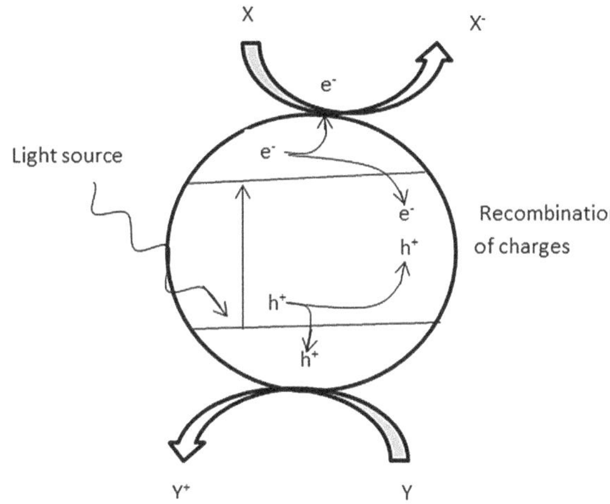

Fig. 7.1 Basic mechanism of photo-catalysis

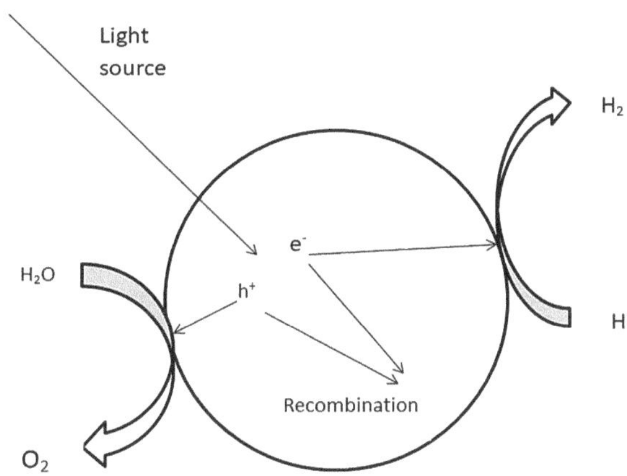

Fig. 7.2 Water splitting reaction using photo-catalysis

From Fig. 7.1, it is clear that the electron absorbs photon and gets excited to jump in the conduction band. So, electron is there in the conduction band, and a hole is present in the valence band. Electron in the conduction band is reducing *X* to X^{-}, and hole is oxidizing *Y* to $Y^{+\cdot}$

From Fig. 7.2, it is understandable that after absorbing photon, electron–hole pair has been generated. The generated electron in the conduction band migrates to combine with water and reduces it to hydrogen, whereas the hole present in the valence band oxidizes the water to oxygen. It is evident that oxidation and reduction both are occurring simultaneously. So the potential corresponding to the conduction band should be more negative, and the potential corresponding to the valence band should be more positive in comparison to the oxidation potential of the hydrolysis reaction to carry out this redox reaction. This phenomenon is known as water splitting reaction using photo-catalysis. Further, co-catalysts are often used which are the materials used with photo-catalyst to trap hole or electron so that

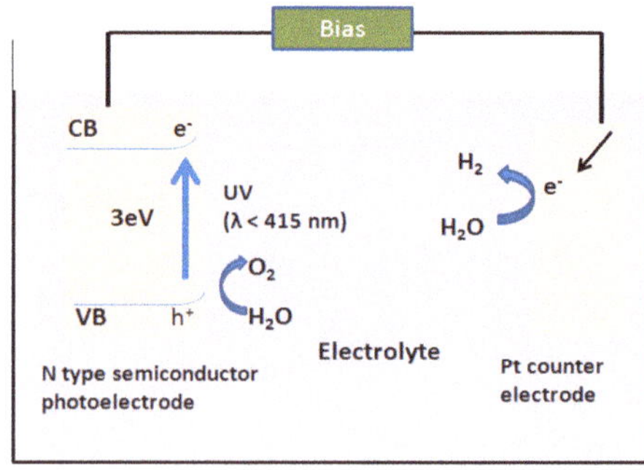

Fig. 7.3 Schematic of the experimental set-up used for water splitting purpose

recombination can be delayed or can be stopped and it can also alter the band gap, for example, graphene oxide, nickel, copper, manganese. First time, Honda–Fujishima water splitting using titanium oxide photo-electrode was reported in 1972 (Tryk et al. 2000).

As it is evident from the Fig. 7.3, it is a normal electro-chemical cell which comprises of two electrodes merged in an electrolyte. Herein, titanium oxide (TiO_2) semiconductor acts as cathode and platinum as anode. Upon exposing it to the sunlight having energy more than 3 eV, an electron–hole pair is generated. Electrons get conducted to anode and reduce the water to hydrogen. The hole present in the valence band of TiO_2 electrode oxidizes the water to oxygen (Tryk et al. 2000; Schneider et al. 2014).

Photo-catalysis based on functionalized nano-materials has now become attractive emerging method in water treatment. There are two distinct sources of energy extensively utilized in the photo-catalytic applications in water treatment: solar energy and artificial ultraviolet (UV) light. Both of these types can be used at the ambient temperature for degrading both organic and inorganic matter as well as microbial pollutants in water.

Out of these two energy sources mentioned, solar energy is naturally available for us. Therefore, solar photo-catalysis technology is inexpensive, environmental welcoming and universally acceptable. It is also appropriate for less developed countries or places with no access to electricity.

7.1.2 Working Principle of Photo-catalysis

The working theory of photo-catalysis is easy to understand. A catalyst exploits ultraviolet radiation from sunlight and utilizes the energy to break down substances present in the water solution. Photo-catalysis process can be used to break down a wide variety of organic materials including textile dyes, micro-organisms, inorganic substances so and so forth. Photo-catalysis coupled with other techniques, viz. filtration, precipitation, is often utilized for the removal of toxic and hazardous metals (such as mercury). Photo-catalysis phenomenon is also explored with the assistance of functionalized nanoparticles as catalysts. Attempts have been made to manufacture nanoparticles having superhydrophobic properties so that the self-cleaning nature could be incorporated in the photo-catalyst which in addition to water purification can be utilized for reducing air pollution as well (Schneider et al. 2014; Gupta et al. 2015a).

7.1.3 Photo-catalytic Study

The schematic of the photo-catalytic mechanism undergone in the presence of semiconducting structures is shown in Fig. 7.4. The equations mentioned below are generally utilized as a model reaction for the representation of photo-catalytic phenomenon with various metal oxide nanoparticles (TiO_2, ZnO, etc.). When semiconductor material gets energy equivalent to their excitation wavelength, electron present in the valence band gets excited. On excitation, electron reaches to conduction band creating void space in the form of hole in the valence band. The induced hole in the valence band is helpful in oxidizing the H_2O molecules to a radical species. It further assists in the oxidation of an organic compound present in the solution. Electron reached on conduction band with the help of solar energy/UV irradiation is carried by oxygen molecule which acts as electron acceptor in this case. Additionally, a superoxide radical is formulated when it is occupied by O_2. Further, it reacts with H_2O to form hydrogen peroxide $(H_2O_2)^-$ which on further oxidation generates OH· free radical. These OH· radicals attack the reactant species (say, dye molecule) to degrade them into CO_2, water and minerals. In a nut shell, electron–hole pair is formed by excitation and that assists in the creation of hydroxyl radicals. These are ultimately utilized in oxidizing any (in most cases, organic or inorganic) species present in the solution to be remediated. Following is the reaction happened in the photo-catalytic system curing with ZnO as photo-catalytic material (Schneider et al. 2014; Gupta et al. 2015a).

$$\mathrm{Zno} + hv \rightarrow e^-_{CB} + h^+_{vB}$$

$$h^+_{vB} + \mathrm{H_2O} \rightarrow \mathrm{H}^+ + \mathrm{OH}^{\cdot}$$

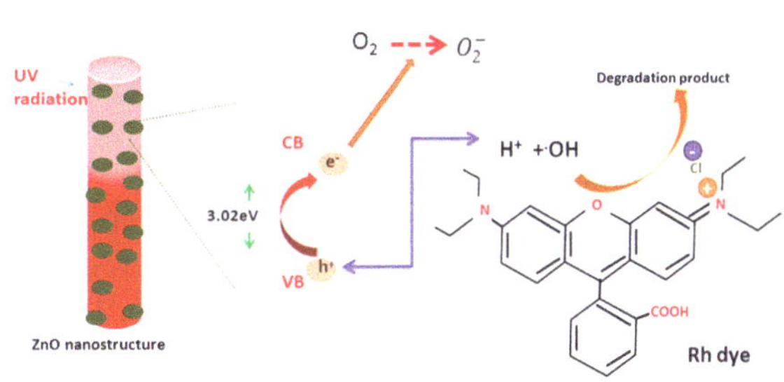

Fig. 7.4 Schematic diagram showing the photo-catalytic phenomenon (Gupta et al. 2015a)

$$e_{CB}^{-} + O_2 \rightarrow O_2^{-}$$

$$O_2^{\cdot -} + OH^{\cdot} + \text{Rh dye} \rightarrow \text{degraded product}$$

7.1.4 Basic Criterion for Selection of Material as Photo-catalyst

The most basic criterion for the selection of the photo-catalyst is that the material must be having reasonable amount of energy gap between the conduction and the valence band. Conduction band is the band of energy levels in which the electrons are having high mobility and the electrical conductivity is caused by them. Valence band is the band which is having low energy level than conduction band. It is the band in which the electrons are present at absolute zero temperature. The energy gap should be such that it should utilize maximum of the sunlight for the excitation of the electron from one (valence) band to other (conduction band). The quantum efficiency of the material should also be maximum.

7.1.5 Types of Photo-catalysts

Among all the photo-catalysts, TiO_2 is the most familiar one. Another material explored is ZnO which could be a practical substitute for some applications.

The efficiency of the photo-catalyst can be enhanced when nano-phase of the same could be explored. As an example, nano-titanium dioxide (nano-TiO_2) can be understood here in this case. Nano-sized particles are preferred to bulk TiO_2 because they are significantly more reactive than larger particles due to their larger surface area. This is chemically stable and has greater abilities to lose molecular bonds which will lead to degradation. Additionally, it is profuse and inexpensive. On using advanced oxidative processes (AOPs), ZnO can serve as a good alternative of titanium dioxide as a photo-catalyst in water treatment. As another photo-catalyst, the degradation of textile dyes can be photo-catalysed by ZnO.

For the first time, the work on the semiconductor as catalyst was demonstrated by Fujishima and Honda. Use of TiO_2 as photo-catalyst for the remediation of water is much known. It has the capacity to disinfect a broad variety of organic pollutants into CO_2 and inorganic ions which are not harmful for environment (Lazar et al. 2012). Nano-crystalline titanium dioxide is a multitasking semiconductor. It works as a catalyst which is used for the treatment of polluted water and air (Malato et al. 2009). It has not only been utilized as a catalyst used for the hydrogen production by splitting water technique but also, it has found immense utility in solar cell applications. The importance of titanium dioxide as catalyst can be understood from the fact that the photo-catalysis and hydrophobicity which are two completely different aspects can occur at the same time on the surface of titanium dioxide (Fujishima et al. 2007). It has been accomplished with the basic requirement for work as catalyst as it is having reasonable amount of gap between the valence band and conduction band and high oxidation potential (Linsebigler et al. 1995).

TiO_2 mainly occurs in three different morphologies anatase, rutile and brookite phase. Rutile is the morphology which is most stable and abundant. Anatase and the brookite can also be converted into rutile upon calcination (Pelaez et al. 2012). The band gap for the rutile, anatase and the brookite is 3.0, 3.2 and 3.2 eV. The semiconductor TiO_2 induces a series of reductive and oxidative reactions on its own surface. The distinct lone pair of electrons in its outer orbital mainly contributes the process. When the photon energy illuminated on the surface of TiO_2 exceeds the band gap energy of TiO_2, the lone pair of electron gets photo-excited to the empty conduction band. The light wavelength for such photon energy usually corresponds to wavelength less than 400 nm. This photon excitation creates electron–hole pair.

One of the major advantages of using titanium dioxide is the abundant availability of the material. Instead of being economical, its generated holes are characterized as highly oxidizing with redox potential of 2.53 V with respect to standard hydrogen electrode. The photo-generated electrons are having redox potential of −0.52 V which can easily reduce di-oxygen into super-oxides. Photo-catalysis using titanium dioxide as photo-catalyst does not depend on the intensity of the incident photon but on the energy.

Drawback of using TiO_2 is that pure nano-TiO_2 is activated by the wavelength range corresponding to UV light. Therefore, for indoor applications, it is necessary for researchers to develop some catalysts that can be activated by visible light spectrum. As far as its impact on health and environment is concerned, TiO_2 is inert and has been used in bulk form in various materials without negative impacts.

7.1.6 Method of Preparation of Photo-catalyst Used for Water Remediation

Since from the above excerpts, it has been clear that different forms of titanium dioxide and the zinc oxide material are mainly used for the photo-catalysis phenomenon for water remediation. Zinc oxide has its own advantage over titanium dioxide as it has wide band gap so that it can utilize maximum amount of sunlight as it can also utilize the light of visible range which is almost 45% of solar irradiation.

Due to ease in fabrication, zinc oxide is also used for water remediation. There are many techniques for the fabrication of the zinc oxide nano-structure. There is a method reported in Gupta et al. (2015a). In this method, some desired molar concentration of zinc acetate is dissolved in the isopropyl alcohol (IPA) by ultrasonicating it for sufficient time so that all the particles of zinc acetate could get

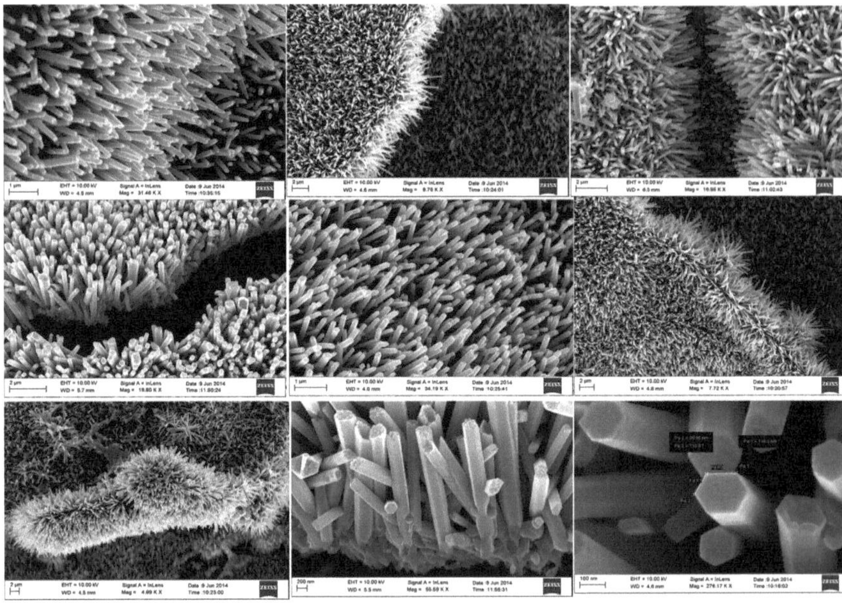

Fig. 7.5 FESEM image of ZnO nano-structures used for photo-catalysis (Gupta et al. 2015b). Reproduced with kind permission

reduced in size. ZnO solution is coated on the silicon substrate. After this step, prepared sample is provided thermal treatment at 90 °C for a day long incubation in the solution of zinc nitrate and hexamethylene tetra amine (HMTA) to provide the atmosphere of zinc and oxide ions for the growth vertically standing structures (Gupta et al. 2015b). Figure 7.5 shows the field emission scanning electron microscope (FESEM) micrographs of the ZnO nano-structures prepared by the above discussed method.

7.2 Basics of Photo-catalytic Reactor Used for Water Purification

Photo-catalytic reactor is a device which is used to make the phenomenon of photo-catalysis happens. To use the phenomenon of photo-catalysis for water treatment on industrial level, there must be the photo-catalytic reactor. It mainly consists of a fixed wall which provides a base for the photo-catalysts or the surface containing photo-catalysts and makes them immobilized. Photo-catalytic reactor utilizes UV visible light from sun, and it has its own source of producing light so that the phenomenon of photo-catalysis can take place even in the night, in the rainy season or cold season when the intensity of sunlight is low.

The design of the photo-catalytic reactor is influenced by many factors as the distribution of pollutants and photo-catalyst inside the bulk, the irradiation properties, rate of mass transfer of pollutants and the reaction kinetics (Mccullagh et al. 2011). The design of the photo-reactor should be such that it should provide high surface area for the stationary photo-catalyst so that it can get exposed to sunlight and can utilize maximum of it. High surface area of the photo-catalyst will also assist in enhancing the mass transfer rate and increase the rate of adsorption to enhance the photo-catalytic activity. It must be designed in such a way that it should consume less energy to perform the photo-catalytic activity.

7.2.1 Basic Principles of Reactor Design

Reactor design mainly focuses on high surface area of contact so that it can come in the contact with the more organic pollutants and neutralize them effectively. It will also enhance the exposure of the photo-catalysts to sunlight so that it can show better photo-catalytic activity. The photo-catalysts which are immobilized on the surface or fixed wall inside the photo-catalytic reactor should not be washed away or mobilized due to the treatment of the continuous flow of the water. The radiation field in the reactor should be such that it should illuminate maximum of the photo-catalyst. The amount of water treated per unit area of photo-catalyst should be maximum.

7.2.2 Different Types of the Photo-catalytic Reactor

Over the time, different types of photo-catalytic reactor have been developed by the researcher to increase the effectiveness of the system by modifying its shape and size and using different type of fixed wall/surface for the immobilization of the photo-catalysts. The modifications were mainly focussed on increasing the area of contact so that the rate of mass transfer and the quantum yield can be enhanced to achieve better photo-catalytic activity.

7.2.2.1 Slurry Reactor System

It is a type of reactor system which was used earlier for the degradation of the organic compound present in the water using the phenomenon of photo-catalysis. It was used mainly using titanium dioxide as photo-catalyst. In this type of photo-catalytic reactor, the photo-catalyst was directly sprinkled into the water which needs to be treated. Due to its small size of the titanium dioxide, it gets suspended in the water. Upon illumination, the photo-catalysts start to show the photo-catalytic effect by degrading the organic pollutants. Due to high surface area of the material present inside the reactor system, rate of mass transfer as well as rate of photo-catalytic activity get enhanced as more surface area is being exposed to sunlight. There are some difficulties associated with this type of reactor as it cannot treat the continuous flow of water and the particles of photo-catalysts which need to separate out from the treated water was very expensive and time-consuming process.

The use of H-titanate nano-fibre for the degradation of the Congo red is investigated by Nan Chong et al. in annular slurry reactor system. It is aimed to find the solution for easy separation of the photo-catalyst nanoparticles. But it also limits the amount of sunlight for irradiation due to the shielding by nano-fibres (Chong et al. 2009). To get rid of this problem, the other method is to use a transparent support on which the catalyst needs to be attached. It resolves the problem of illumination and separation of catalyst since it is attached to the support. Drawback of this system is that it slows down the rate of adsorption.

In 2013, Morgan Adams suggested the method of effective mixing to increase the rate of degradation of pollutants in the slurry reactor. In this method, it has been found that the effective mixing increases the rate of mass transport and resulted in the reduction of organic pollutant by 85% (McCullagh et al. 2010).

7.2.2.2 Flat Plate Reactor

Flat plate reactors are the simplest in design as it contains a flat plate on which the photo-catalysts are adhered. For the phenomenon of photo-catalysis, it is placed in a position such that water can be flowed down easily with the help of gravity. The speed of water flowing down can be adjusted by providing suitable inclination to the plate. The flat plate which contains photo-catalysts is also illuminated with light to activate the photo-catalysis phenomenon and to degrade the organic pollutant of the water flowing down through the plate. Due to the large surface area of the plate and the photo-catalysts, there is a large surface area for the activity of photo-catalysis. It is also benefitted with uniform distribution of light as it also causes uniform degradation of the pollutants. As due to the non-uniformity of irradiation, the mass transfer of the organic pollutants can be limited. It also has the limitation that only a thin layer of water comes in contact with the photo-catalysts. Schematic of flat plate photo-catalytic reactor is shown in the Fig. 7.6.

A flat plate photo-catalytic reactor developed by Nogueira and Hardim using titanium dioxide as photo-catalyst adhering on the glass plate and the inclination of the plate was 22°. From this experiment, it was seen that the degradation of organic pollutant was 95.8% at 22° slope and 89% at 25° slope (Nogueira and Jardim 1996).

A flat plate reactor was designed by Adams for the removal of organic pollutants from water. He developed a multilayer flat plate reactor which was based on the design developed by the Nogueira and Hardin. He conducted experiment on two different flat plate made up of polymethylemethacrylate (PMMA) and titanium. The advantage of using these as plates was that PMMA is transparent to UV light which

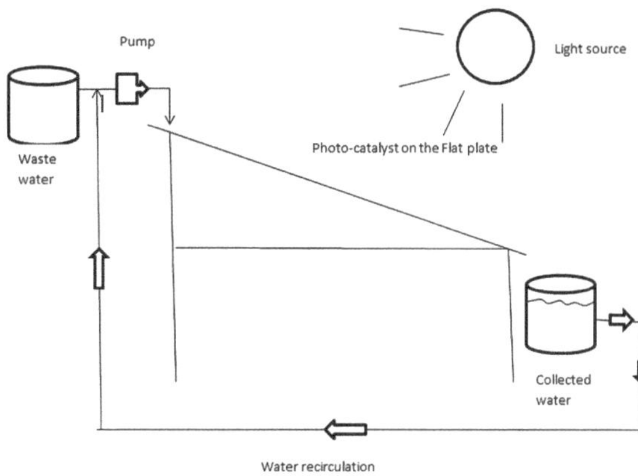

Fig. 7.6 Schematic diagram of flat plate photo-catalytic reactor (Nogueira and Jardim 1996)

increases the illumination and increases the photo-catalytic activity, and the titanium produces titanium dioxide on oxidation which act as the photo-catalyst. The plate made up of PMMA gave better result than the titanium plate due to the better illumination (Adams et al. 2008). It was also observed that the efficiency is increasing with the decreasing angle of the slope.

7.2.2.3 Rotating Disc Reactor

It is a type of photo-catalytic reactor which is used for the degradation of organic pollutants by fixing the photo-catalyst on the rotating disc. Rotating disc provides a support to the photo-catalyst. It can be used for the treatment of water in continuous flow manner or stationary manner. The rate of degradation can be controlled by the intensity of light, exposed surface area of the photo-catalyst, velocity of the disc (Dionysiou et al. 2000). Figure 7.7 illustrates the various parts of a rotating disc reactor.

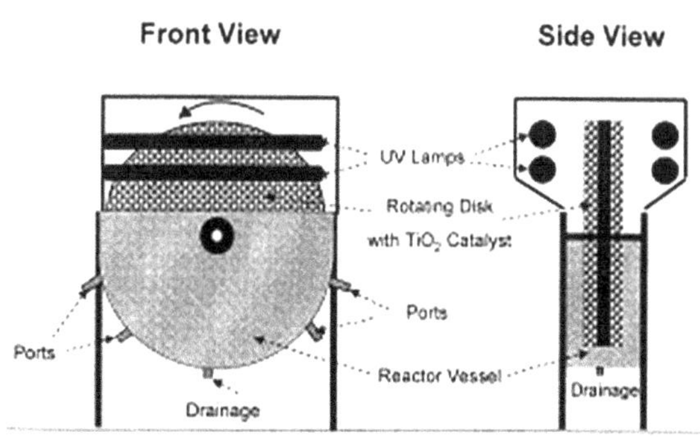

Fig. 7.7 Rotating disc photo-catalytic reactor (Dionysiou et al. 2000). Reproduced with kind permission

7.2.2.4 Honeycomb Monolith Reactor

Monoliths are made up of different type of channels either interconnected or separate in a single block of material. These types of photo-catalytic reactors are of honeycomb-shaped monolith. It contains a series of small channels which provide a path to the flow of water. The photo-catalysts are coated on the interior surface of the comb-like structure. The main feature of this type of photo-catalytic reactor is that it provides very high surface to volume ratio and it causes very less pressure loss (Hossain and Raupp 1999; Tahir and Tahir 2016; Taranto et al. 2009).

7.2.2.5 Fluidized Bed Reactor

The polluted water passes through the catalyst at high speed which causes the suspension of the solid particles of photo-catalyst. The behaviour of particles becomes similar to fluid. This type of photo-catalytic reactor supports photo-catalysts particles on a fluidized bed with quartz. Due to the fixing of the catalyst particles on the fluidized sand bed, there is a very less chance of flow of catalyst particles with continuous flow of water. Quartz has been selected for the support of photo-catalysts due to its UV conducting nature (Kanki and 2005; Chiovetta et al. 2001).

7.2.2.6 Optical Fibre Reactor

In this photo-catalytic reactor, the photo-catalyst has been immobilized on the optical fibre than other fixed surface. Due to the immobilization of the photo-catalyst on the optical fibre, it provides the more uniform distribution of the photo-catalyst throughout the volume. The photo-catalysts are in the direct contact with the light emitting source, so the chance of the loss of the light is minimum. The performance of this type of reactor is highly influenced by the coating thickness of catalyst on the fibre, their length and on the diameter of the fibre itself (Danion et al. 2004).

7.2.2.7 Tube Light Reactor

This modelled reactor contains fluorescent tube of extremely narrow diameter emitting light of required wavelength. Due to narrow diameter of the tube, it can be placed in large number. The photo-catalyst has been deposited on the outer surface of the narrow diameter light emitting tube. Figure 7.8 shows the tube light photo-catalytic reactor.

The advantage of this method over others is that in this method, the light is directly falling on the catalyst, so like other method light does not need to pass through the reactant and the product and then to the catalyst. It also does not have to

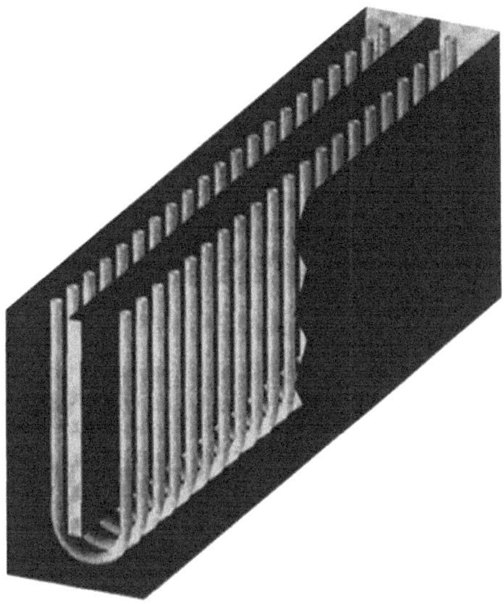

Fig. 7.8 Schematic of tube light photo-catalytic reactor (Ray and Beenackers 1998). Reproduced with permission

pass through the bulk of wastewater which causes loss of irradiation. In this way, it enhances the photo-catalytic activity of the catalyst (Ray and Beenackers 1998).

7.3 Life Cycle Assessment of Photo-catalytic Water Treatment Process

Life cycle assessment of any process is the detailed study of the effect of a particular process on the environment. Using life cycle assessment (LCA), impact of the material on the environment, its use and way of its discharging, etc., are studied. Assessment may include the studies related to the role of material on global warming, its consequence on ozone layer, its toxicity so and so forth. It is also said cradle to grave analysis of the process (Yadav and Mishra 2013). It is a tool which is used to check the effect of the process or the product on the environment (Guinée et al. 2000). Water treatment process through photo-catalysis makes use of the photon provided by the sun, of required energy and wavelength (Legrini et al. 1993). Solar energy is a renewable and clean source of energy. The phenomenon of photo-catalysis is mainly driven in the presence of sunlight, so no need of external energy source is needed. There are many types of photo-catalysts which can be used for photo-catalysis:

1. Titanium oxide-based photo-catalysis (Fujishima et al. 2000)
2. Zinc oxide-based photo-catalysis (Kołodziejczak-Radzimska and Jesionowski 2014)

The zinc oxide as photo-catalyst for water remediation absorbs sunlight, which is a clean and renewable source of energy. It uses the photon to excite the electron from valence band to conduction band. The excited electron in the conduction band is used in reduction of oxygen to produce oxygen radical of oxygen and the hole generated in the conduction band combines with hydroxyl to generate free radical of hydroxyl which further reacts with organic pollutants to degrade them. Since the process is clean and very efficient in term of energy, there are various risk associated with the nanoparticle.

7.4 Challenges of Water Remediation Using Photo-catalysis

The one of challenges in using solar-based photo-catalysis for water remediation is that it needs sunlight for the remediation of water. Its use is limited for winter season in the cold areas. It can work only in the particular area where the sun shines on regular basis. Due to the absence of the sunlight, it will be difficult to operate it in the night time, winter season and in the rainy season.

Another challenge lies in the choice of material. Material chosen for photo-catalysis must be chosen very carefully keeping in mind that the band gap, i.e. the gap between valence band and the conduction band should be in the range of energy produced by visible range of light as the sunlight contains almost 45% visible light (Kumar and Devi 2011). So it will ensure the maximum utilization of the sunlight. For example, due to wide band of titanium oxide, sunlight of high energy is needed for exciting the electron from valence band to conduction band (Fujishima et al. 2007). The band gap of titanium oxide is almost equal to 3.2 eV (Morikawa et al. 2001), which means it needs sunlight of wavelength less than 400 nm. Sunlight having energy less than 3.2 eV will be unable to excite the electron to conduction band. However, most of the sunlight which falls in the range of 400–700 nm range is of no use for this photo-catalysis process (Herrmann 1999).

The material should be selected in such a manner that instead of having proper band gap, it should have maximum efficiency to excite electron from valence band to conduction band on exposure to the photons (Hou et al. 2006). The affinity of organic pollutants towards photo-catalyst is also a major factor which determines the rate of the photo-catalytic reaction. Low affinity between the pollutants and the photo-catalyst cause them not to come in contact readily. As a result, it causes the decrease in the rate of the adsorption. So to increase the efficiency, it is necessary that the organic pollutants should readily come in contact with photo-catalyst for high rate of mass transfer and adsorption. The contact surface area of photo-catalyst exposing to light as well as the pollutants must be high, and there must be the affinity of the organic pollutants towards the photo-catalyst. As an example, it does not happen in the case with the titanium oxide photo-catalyst due to its poor affinity with organic pollutants (Mahlambi et al. 2015).

One of the major challenges of the photo-catalytic reaction is to increase the time of recombination of the electron and holes. Recombination of electrons and holes causes the releases of energy, and it is not a desirable phenomenon. It is desired that the recombination must be delayed as much as possible so that excited electron gets more time to reduce the ion in the water. So there is a need of suitable material for increasing quantum efficiency by electron scavenging for the immediate removal of the generated electron and to increase the time of recombination and to improve the photo-catalytic performance of the photo-catalyst (Ohtani 2013). Apart from that, the instability of the nanoparticles is the major concern which needs to be addressed. Due to the instability of the nanoparticles, the chance of forming aggregates of titanium oxide nanoparticle by coagulation is very often which decreases the area of contact for both the adsorption as well as the catalytic activity. So to decrease the tendency of the aggregation is a big challenge. It has been addressed by using the supported catalysis which has reduced the problem of aggregate formation, and it has been useful in continuous flow water remediation. But at the same time, it also reduces the area of catalyst for the exposure, to sunlight for the generation of the electron–hole pair and the pollutants for adsorption. The use of material for the support also increases the cost of the system. Although this method of remediation has been a great success for water remediation, still this method has not been reported with 100% efficiency, i.e. it has not been able to remove the organic pollutants completely.

The efficiency of the photo-catalysts to degrade the organic pollutants decreases with time. Its ability to degrade the organic pollutants reduces with time. The rate of degradation of organic pollutants also decreases with increase in initial concentration of the organic pollutants (Kim et al. 2006). By manipulating the amount of the oxygen defect, the carrier concentration of zinc oxide can be controlled. The causes of oxygen defect in the zinc oxide crystal are due to the escaping of oxygen atom from the lattice site due to the low partial pressure and high annealing temperature. The oxygen defect is maximum in the case of high aspect structures, viz. rod/wire-shaped structure. Size and shape of the structures change based on the fabrication environment and other process variables provided. These variables may include synthesis temperature, pressure and chemicals used for preparation. Choice of the synthesis procedure, temperature and pressure conditions will be responsible for the crystal defects. These defects play very important role in the photo-catalytic reaction (Hamid et al. 2017).

With the passage of time, a lot of progress has been made, but there is no as such precise technique or method which has been developed for the removal of nanoparticles suspended in the wastewater during water treatment. Due to its small size, it creates problems for environment as well as for human health which is necessary to be addressed.

7.5 Conclusion and Future Scope

Researchers all over the world are looking forward for the cheaper solution to global challenges, viz. water remediation, etc. Use of solar energy for water remediation is the best possible solution in this era. Photo-catalysis-based water remediation is the well-explored phenomenon which not only can be carried out with the help of natural energy source, i.e. solar energy but also a reliable purification technique.

Future scope lies in the exploration of methods for increasing the photo-catalytic efficiency. This can be achieved by the chemical modification of the photo-catalysts, exploration of high surface nano-materials as well as manipulating the band gap by doping noble metals to the semiconducting materials.

References

Adams M, Campbell I, Robertson PKJ (2008) Novel photocatalytic reactor development for removal of hydrocarbons from water. Int J Photoenergy. 2008

Chiovetta MG, Romero RL, Cassano AE (2001) Modeling of a fluidized-bed photocatalytic reactor for water pollution abatement. Chem Eng Sci 56(4):1631–1638

Chong MN, Jin B, Zhu HY, Chow CWK, Saint C (2009) Application of H-titanate nano-fibers for degradation of congo red in an annular slurry photo-reactor. Chem Eng J 150(1):49–54. ISSN 1385-8947

Danion A, Bordes C, Disdier J, Gauvrit J-Y, Guillard C, Lantéri P, Jaffrezic-Renault N (2004) Optimization of a single TiO-coated optical fiber reactor using experimental design. J Photochem Photobiol A Chem 168(3):161–167

Dionysiou DD, Balasubramanian G, Suidan MT, Khodadoust AP, Baudin I, Laîné J-M (2000) Rotating disk photocatalytic reactor: development, characterization, and evaluation for the destruction of organic pollutants in water. Water Res 34(11): 2927–2940. ISSN 0043-1354

Fujishima A, Rao TN, Tryk DA (2000) Titanium dioxide photocatalysis. J Photochem Photobiol C: Photochem Rev. 1(1):1–21 ISSN 1389-5567

Fujishima A, Zhang X, Tryk DA (2007) Heterogeneous photocatalysis: From water photolysis to applications in environmental cleanup. Int J Hydrogen Energy 32(14):2664–2672

Guinée JB, Gorree M, Heijungs R, Huppes G, Kleijn R, Udo De Haes HA, Van Der Voet E, Wrisberg MN (2000) Life cycle assessment. An operational guide to ISO standards centre of environmental science. Leiden University (CML), The Netherlands

Gupta A, Mondal K, Sharma A, Bhattacharya S (2015a) Superhydrophobic polymethylsilsesquioxane pinned one dimensional ZnO nanostructures for water remediation through photo-catalysis. RSC Adv 5(57):45897–45907

Gupta A, Saurav JR, Bhattacharya S (2015b) Solar light based degradation of organic pollutants using ZnO nanobrushes for water filtration. RSC Adv 5(87):71472–71481

Hamid SBA, Teh SJ, Lai CW (2017) Photocatalytic water oxidation on ZnO: a review. Catalysts 7 (3):93

Herrmann JM (1999) Heterogeneous photocatalysis: fundamentals and applications to the removal of various types of aqueous pollutants. Catalysis Today 53(1):115–129

Hossain MM, Raupp GB (1999) Polychromatic radiation field model for a honeycomb monolith photocatalytic reactor. Chem Eng Sci 54(15):3027–3034

Hou X, Wu X, Liu A (2006) Studies on photocatalytic activity of Ag/TiO_2 films. Front Chem China 1(4):402–407

Kanki T, et al. (2005) Water purification in a fluidized bed photocatalytic reactor using TiO_2-coated ceramic particles. Chem Eng J 108(1):155–160

Kim TS, Kim JK, Choi K, Stenstrom MK, Zoh KD (2006) Degradation mechanism and the toxicity assessment in TiO photocatalysis and photolysis of parathion. Chemosphere 62 (6):926–933 ISSN 0045-6535

Kołodziejczak-Radzimska A, Jesionowski T (2014) Zinc oxide—from synthesis to application: a review. Materials 7(4):2833–2881

Kumar SG, Devi LG (2011) Review on Modified TiO2 Photocatalysis under UV/visible light : selected results and related mechanisms on interfacial charge carrier Transfer Dynamics.13211–13241

Lazar M, Varghese S, Nair S (2012) Photocatalytic water treatment by titanium dioxide: recent updates. Catalysts 2(4):572–601

Lee KM, Lai CW, Ngai KS, Juan JC (2016) Recent developments of zinc oxide based photocatalyst in water treatment technology: a review. Water Res 88:428–448

Legrini O, Oliveros E, Braun AM (1993) Photochemical processes for water treatment. Chem Rev 93:671–698

Linsebigler AL, Lu G, Yates JT (1995) Photocatalysis on TiO_2 Surfaces: principles mechanisms, and selected results. Chem Rev 95(3):735–758

Mahlambi MM, Ngila CJ, Mamba BB (2015) Recent developments in environmental photocatalytic degradation of organic pollutants: the case of Titanium Dioxide nanoparticles—a review. J Nanomaterials, vol. 2015

Malato S, Fernanndez-Ib P, Maldonado MI, Blanco J, Gernjak W (2009) Decontamination and disinfection of water by solar photocatalysis: Recent overview and trends. Catal Today 147 (1):1–59

McCullagh C, Robertson PKJ, Adams M, Pollard PM, Mohammed A (2010) Development of a slurry continuous flow reactor for photocatalytic treatment of industrial waste water. J Photochem Photobiol A: Chem 211 (1):42–46. ISSN 1010-6030

Mccullagh C, Skillen N, Adams M, Robertson PKJ (2011) Photocatalytic reactors for environmental remediation: a review. J Chem Technol Biotechnol 86(8):1002–1017

Morikawa T et al (2001) Band-gap narrowing of titanium dioxide by nitrogen doping. Japan J Appl Phy 40(6A):L561

Newton MD, Sumi H, Skourtis SS, Beratan DN, Vanmaekelbergh D, Hammes-Schiffer S, Piotrowiak P, Endicott JF, Fox MA, Andersson AM, Schmehl RH (2001) Electron transfer in chemistry

Nogueira RF, Jardim WF (1996) TiO_2-fixed-bed reactor for water decontamination using solar light. Sol Energy 56(5):471–477

Ohtani B (2013) Titania photocatalysis beyond recombination: a critical review. Catalysts 3:942–953

Pelaez M et al (2012) A review on the visible light active titanium dioxide photocatalysts for environmental applications. Appl Catal B Environ 125:331–349

Ray AK, Beenackers AACM (1998) Novel photocatalytic reactor for water purification. AIChE J 44:477–483. https://doi.org/10.1002/aic.690440224

RSC Advancing the Chemical Sciences. TiO_2: Uses of titanium dioxide. Wolfson Found, no. 207890

Schneider J, Matsuoka M, Takeuchi M, Zhang J, Horiuchi Y, Anpo M, Bahnemann DW (2014) Understanding TiO_2 photocatalysis: mechanisms and materials. Chem Rev 114(19):9919–9986

Tahir M, Tahir B (2016) Dynamic photocatalytic reduction of CO_2 to CO in a honeycomb monolith reactor loaded with Cu and N doped TiO_2 nanocatalysts. Appl Surf Sci 30(377):244–252

Taranto J, Frochot D, Pichat P, (2009) Photocatalytic air purification: comparative efficacy and pressure drop of a TiO_2-coated thin mesh and a honeycomb monolith at high air velocities using a 0.4 m close-loop reactor. Sep Purif Technol 67(2):187–193. ISSN 1383-5866

Tryk DA, Fujishima A, Honda K (2000) Recent topics in photoelectrochemistry: achievements and future prospects. Electrochim Acta 45(15):2363–2376

Yadav SK, Mishra GC (2013) Environmental Life Cycle Assessment Framework for Sukker Production (Raw Sugar Production)

Chapter 8
Water Splitting by Using Electrochemical Properties of Material

Deepak Kumar Tiwari, Apurba Ray, Atanu Roy and Sachindranath Das

Abstract The present era of ramping energy demands, its storage considerations coupled with environmental contentions being caused by the perishable conventional resources of energy, has set out a clear tone for the research community all across the globe to impart a motivated effort over it. If found, the genuine solution will go a long way to shift the paradigm from the current fossil-based economy to hydrogen-based one. The article at hand prima facie focuses on one of the most promising solutions to the aforementioned problem namely water splitting—more specifically so by harnessing the electrochemical properties of the material. While many options exist to furnish electricity for carrying out the water splitting but the cost compatibility with the other contemporary candidates poses the real bottleneck. The idea over here is to efficiently utilize the hitherto non-conventional methods of energy generation to efficiently and cost effectively achieves water splitting for the subsequent applications. As regards the current trends in water splitting, recent advances include carrying out water splitting by incorporating the visible portion alone from the available light. The resulting hydrogen may then subsequently be employed as a fuel input for different applications. Routes ranging from using advanced catalysis and even artificial photosynthesis are also being considered.

Keywords Alternative energy resource · Electrocatalysis · HER OER · Photocatalysis · Water splitting

D. K. Tiwari · A. Ray · A. Roy · S. Das (✉)
Department of Instrumentation Science, Jadavpur University, Kolkata 700032, India
e-mail: sachindas15@gmail.com

D. K. Tiwari
Cochin University of Science and Technology, Cochin, Kerala, India

S. Bhattacharya et al. (eds.), *Water Remediation*, Energy, Environment, and Sustainability, https://doi.org/10.1007/978-981-10-7551-3_8

8.1 Introduction

Alternative energy, without any ambiguity, is the need of the hour. We cannot anymore draw endlessly from the fossils and use nature unsustainably like a commodity. One of the most crucial questions mankind asks itself today is about the ways in which it can pass on the planet in a cleaner and greener shape than it received to the coming generations. Reducing carbon footprints while simultaneously being in tandem with our developmental ambitions is the real challenge we must gear up for.

From global warming to ozone hole depletion, from greenhouse gasses emission to animal and plant species becoming extinct—all of them are the manifestations of the same issue. The need of the hour is quite clear viz a naturally congenial but industrially viable solution at the earliest. If emissions continue unchecked, the repercussions may end up spending millions of refugees across the borders, causing the widespread extinction of most species on earth and melt the polar ice caps leaving most of the world's coastal cities entirely submerged underneath the sea. This naturally calls out for a major technological breakthrough footed on the best available tools and naturally congenial behaviors currently available at our disposal. If waterbodies continue to rise unchecked, we are looking at a rise of several feet of the water level which will cause a widespread abandon of some of the major cities we have built resulting us to adapt to an entirely altered set of demographic situations. What is evident from past climatic change patterns is that every time the level of CO_2 in the atmosphere has suffered a rise, the earth warms up resulting in the ice caps to melt and the ocean water rises. Let us face the fact that we are pumping green house gasses into the atmosphere at a rate history has never ever seen before. As a result, climatic warming is making heat waves more frequent causing increased drought scenarios and heavier rain storm spells simultaneously in different parts of the earth. What one is rightly made to remember is the famous quote by the US President J.F. Kennedy saying, "If not us, who? If not now, when?" and only to supplement it by asking If not Earth, Where? Considering our perpetual record of ignorance to what the scientists have been warning us for over two decades, now invariably is the time to put our best foot forward.

8.1.1 Global Co-operational Endeavors

The first ever serious effort to provide a global platform to any national level congregation came under the aegis of World Meteorological Organization (WMO) and United Nations Environment Program (UNEP). It is Intergovernmental Panel on Climate Change (IPCC) in the year 1988 followed by its assessment reports (the ARs) further supplemented by the Earth Summit of Rio De Janerio and the United Nations Framework Convention on Climate Change, paved the way for the historic Kyoto Protocol.

A typical curve between the biocapacity of the earth versus our ecological footprint gives an ecological overshoot indicative of a ramping increment of our demand vis-à-vis the environmental supply capacity. One of IPCC's most recent assessment report, the AR had fixed the quota of CO_2 from the period of Industrial revolution till the completion of the twenty-first century to 2900 Gigatons, of which we already have surpassed two-thirds the quantity. Adding to the disdain further is the extrapolation that at this rate we would finish the remaining one-third in a matter of around thirty years. The anthropogenic greenhouse emission has increased many percentages owing to the increased use of fossil fuels, deforestation, and the boom in the meat and the livestock industries. The AR also indicates a decrement over the staple crops of wheat, rice, and maize all across the tropical and temperate parts world which should be taken as an indicator of the increase in the per capita cost of living thereby imparting an economic angle to the problem too. As can be inferred from the fifth AR again, a whole quarter of the total anthropogenic greenhouse gas emission is attributed to the electricity and heat production sector. Around another quarter (24%) is caused by the agricultural and other land uses leaving the industry and the transport sector at 21 and 14%, respectively, only to be followed up by the emissions caused by the buildings and other sources. At this rate, it is predicted that by the end of the twenty-first century, under even a low emission scenario, an average increment of 1.5 °C temperature would be imparted onto the earth. A realistic high emission indicates a rise of 2°, whereas a possible unchecked emission can lead up to a 4.8 °C rise in the average earth temperature. Currently, the present greenhouse gas emission amount is about 50 Gigatons which attributes itself to a 4 °C temperature increment of the earth. To keep this amount in tandem with the stipulated amount of 2° more than the preindustrial level, this should have to be only 41 Gigatons.

Right since its inception, the United Nations Framework Convention on Climate Change (UNCCC) has been arranging a yearly meeting of its members more popularly known as the Conference of Parties (CoP) to serve as a platform for the global and regional powers to assess their collective progress in the shared pursuit of climate change mitigation and adaptation, at the same time chalking out their way forward. Herein, it is pertinent to highlight the role of 10th edition of "Conference of Parties", outcomes of which formulated the famous Kyoto Protocol. Held at the end of 1997, this served as the Cornerstone for all the more economically developed countries or MEDCs to set binding obligations on them for reducing their emission of the major greenhouse gasses. The six gasses it mentioned in its third annexure were CO_2, CH_4, N_2O, SF_6, Perfluorocarbons, and Hydrofluorocarbons. Likewise, the Montreal protocol which had taken place earlier in the year 1989 came into effect to lessen the emission of freeons or the chlorofluorocarbons or the CFCs to replenish the ozone layer hole.

8.1.2 Alternative Energy Resources into the Rescue

Apart from the fossil-based fuels, perhaps the most utilized fuel for energy generation is the nuclear energy which comes with its own repercussions of radiation hazards and uncontrolled nuclear disasters such as the Chernobyl catastrophe. This again reiterates the need to ponder over the alternative fuel resources. Moreover, getting rid of the nuclear waste is quite much of a problem in itself.

This lacuna seems to be promisingly filled by tapping the alternate sources of energy. The best part about them being the input, whether air, water, or sunlight are abundant in nature and are almost ubiquitous. On this front, photovoltaic electricity generation seems to be the forerunner as compared to all the other candidates. Moreover, the striking increment in the efficiency value of perovskite-based solar cells from 3.8 to over 20.1% in a matter of less than a decade is really an achievement (Jeon et al. 2014). Also to be noted is that unlike the second generation or the direct band gap materials based solar cells, the absence of naturally hazardous materials like cadmium make their candidature even stronger. Herein, the recent research effort to incorporate tin instead of lead in perovskite solar cells is further noteworthy.

However, the innate issue with solar energy is its sheer unavailability during the night. Also, the very basis for the shift in research paradigm from the Si-based solar cells to their heterojunction alternatives was the *Shockley–Queisser* limit which mathematically halted their efficiency value to a measly 33.7% (Shockley and Queisser 1961). Even after over four decades of research, the present scenario of the most published solar cells seems to be hovering around this limit only. The high cost per wattage value makes their commercialization an even more uphill task which, combined with less robustness and transport immobility, really leaves a lot of scopes for more research to usher in. The aforementioned factoids clearly negate the dependence of solar energy alone in the pursuit of alternative energy. More or less, the same holds for most of other candidates which quite imperatively need a suitable technological breakthrough to fill the void. Various energy-derived water splitting routes and their combinations are shown in Fig. 8.1.

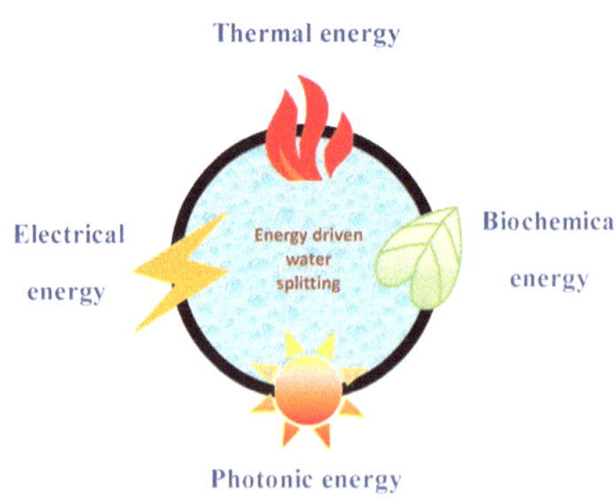

Fig. 8.1 Various energy-derived water splitting routes. Their combination is also a possible route

8.1.3 *Where Does Hydrogen Fit into the Jigsaw?*

We have noted that the non-conventional alternative of harnessing solar energy in its all capacity has got only auxiliary roles to play and can never serve as an alternative fuel in at least its present ambit. This paves the way for the consideration of hydrogen-based fuel alternatives. Hydrogen is, by far, one of the most abundant elements on the earth with the oceans acting as its generous reservoirs. If developed, an ideal splitting technique of water can potentially go a long way to address our energy related concerns in the most environmentally benign way possible.

The challenging part over here is the sizeable amount of energy input it requires to produce enough hydrogen for making the process industrially and hence virtually conceivable. Hence, in this arena, efforts are required. As of now, the major routes to hydrogen production rely heavily on the reforming of hydrocarbons (as in $C_nH_{m+n}\ H_2O \rightleftharpoons (n + m/2)\ H_2 + n\ CO$) or by the production of the syn gas using the coal gasification technique ($3C + O_2 + H_2O \rightarrow H_2 + 3CO$). None of these techniques can be regarded as environment friendly and are used for more than 90% of the industrially driven hydrogen produce. On the other hand, techniques such as electrolysis simply contribute to around 4%.

Due to its ubiquity, water splitting assisted by sunlight is being hailed in high hopes, wherein the third generation solar cells will have important contributions to make. Also the electrocatalytic water splitting techniques which employ materials ranging from noble metals like Pt and Ir (along with their oxides), to different binary and ternary metal cluster complexes (eg. Fe–Ni) have also pretty much substantiated the amount of research. Its combustion to generate energy renders only moisture as the by-product which is not at all problematic from the environmental point of view. This process, on the other hand, releases more than twice energy as the fossil fuels further substantiating the need for a motivated research into the arena of water splitting. The chapter at hand aims to present the latest update in this field. The focus is also on the recent shifts in the paradigm of water splitting and for the sake of brevity; the changes are articulated into the category of either being light assisted (Photocatalytic water splitting) or electrocatalysis driven, wherein a variety of advanced catalytic complexes would be employed into use.

8.2 Photocatalysis

Photocatalysts are semiconductor materials which are known to optimize the rate of reaction. Photocatalysis plays a crucial role in channelizing solar energy, the primary energy source on earth, into a host of different energies (chemical, electrical, etc.). Hence, photocatalysts function as the life sustainers on the earth. At the heart of these life sustainers are the photosynthesis reaction. Mimicking photosynthesis under the laboratory conditions at a requisite rate is termed as the artificial photosynthesis which too comes under the ambit of the water splitting reactions.

$$H_2O^* - 1/2\ O_2(g) + H_2(g); \quad \Delta G = +237\ \text{KJ/Mol} \tag{8.1}$$

Water splitting, under the thermodynamic purview, is an uphill reaction which is evident from its positive Gibb's free energy value of 237 kJ/Mol under the standard conditions.

The major factors which exert influence on photocatalytic materials are as under:

1. Band Gap—This should certainly be in accordance with frequency range pertaining to the light in the visible or at most, the UV range as it is this fraction of light which is of our interest. Light in the UV region (<400 nm) encompasses only around 5% of the total solar energy, whereas the fraction of visible radiation is over 50%, making the visible portion of more significance which is availed using a plethora of techniques as shall be discussed later.
2. Exciton Transport—Ambient carrier migration which can take care of the quick recombination of the generated electron–hole pair is the most sought after attribute of the photocatalysts, failure to which may result in the regeneration of the absorbed photon. Furnishing the photocatalyst in the nanodomain becomes very important here due to the ease experienced by the excitons to reach the active site (Zheng et al. 2015).
3. Surface Area—A large surface area implies a large number of active photocatalytic sites which indeed is the requirement for efficient photochemical reactions.
4. Catalytic Stability—To make the process industrially favorable, the catalyst should be stable enough for being incorporated into repeated cycles of reactions. This increases the industrial affordability of water splitting reactions by many folds.
5. High Crystallinity—This especially coupled with low crystal defects can pose as a recombination inhibiting arrangement due to the appreciable absence of recombination sites.

The aforementioned qualities are best met by the semiconductors (Lee 2011). On one hand, metals, due to their overlapping energy bands allow for only either reduction or oxidation one at a time, the insulators, on the other hand, come with a band gap differential so high that it cannot be mitigated using the light under the UV-visible region. The catalysts need the redox reactions to happen simultaneously, that too preferentially at the extreme surfaces of the same substance which only semiconductors can comply with.

8.2.1 Modus Operandi of the Catalytic Action

The impingement of photons on the catalytic surface results in the generation of the excitons (Walter et al. 2010). Post the generation of the excitons, one has to take care of the surface and volume recombination. Moreover, the carriers generated should be made to migrate as far as possible. The excitons can then be separated simultaneously onto the extreme sides of the same semiconductor and subsequently incorporated for the photocatalysis (Zhang and Yates 2012).

8.2.2 *Minutes of the Photocatalyst-Assisted Water Splitting*

The water couple to be generated on account of the photocatalytic action has the energy difference of 1.23 eV, which the band edges of the catalysts must contain appreciably (Xu et al. 2016). It is required that the potential of the bottom of the conduction band (which corresponds to the oxidation power) should be more negative and the potential of the valence band be more positive with respect to the water couple's potential values. Another determining factor is the extent of the ionic character of the metal oxide bond which when is more, affects the nature of the water splitting reaction substantially.

Mathematically, we have

$$\%\text{Ionic Character} = \left[1 - \exp\left\{-(x_A - x_B)^2/4\right\}\right] \times (100\%) \qquad (8.2)$$

where $x_a - x_b$ denotes the difference between the electronegativity values of the substances *A* and *B*, respectively.

Use of co-catalysts as the exciton traps is also a widely explored area of importance. Metals such as Ni and Pt have been popularly incorporated as the electron traps, whereas metallic oxides such as RuO_2 find a decisive role to play in the hole-trapping process. Such exciton trapping mechanisms increase the recombination time of the charge carriers thereby facilitating the reaction by modifying the catalytic/co-catalytic system's interface.

8.2.3 *Photoelectrolysis of Water*

Photocatalysis, not looking at the water splitting aspects of it, was first reported by Markhama (1955). However, it was not until the path breaking discovery of photoelectrolysis of water, splitting it into its constituent oxygen at a TiO_2 photoelectrode along with hydrogen at the Pt electrode by Fujishima and Honda (1972) that the process of photoelectrolysis began to show promise for the alternative fueling prospects.

The setup typically consists of the photoelectrode on which, the light is to be shinned, typically a semiconductor (e.g., TiO_2) and a metal electrode (e.g., Pt). The choice of the semiconductor should inherently be such that its band gap overlaps with that of the water couple of 1.23 eV along with the aforementioned attributes for water splitting. Normally to impart robustness, the photoelectrode is wrapped around a metal core. Moreover, to address the recombinational bottlenecks, the charge transfer process has been formerly mentioned, should be as fast as possible. A slow recombination process leads to photocorrosion resulting into the loss of impinged photon's energy. The electrons can be taken to transfer from the CB of the semiconductor to that of hydrogen, whereas the migration of the hole from the VB of O_2 to that of the semiconductor.

The electronic and thermodynamic constraints imposed over here too are more or less same as that of the former case, wherein the semiconductor band gap should be more than 1.23 eV. For the band gap exceeding that required by the water couple of 1.23 eV, it is mandatory that the band edges of the semiconductor should overlap that of the water couple.

8.2.4 Some Recently Published Techniques to Enhance Efficiency

The photon absorption can be significantly increased by nanodomain surface modifications coupled with the incorporation of dyes (Chowdhury et al. 2017). This has been found to affect the Incident Photon to Current Conversion Efficiency (IPCE) value several times. Interfacing the back contact suitably with a nanocrystalline material can be of paramount importance. The relation of efficiency values of the catalytic process over the Fermi level changes as a result of the substrate's electron transfer is of importance too (Yin et al. 2016). Doping of suitably wide band gap semiconductors in an attempt to create impurity states can turn out to be important from the electron transfer process' point of view (Chowdhury et al. 2017). The electrons then can attain the intermediately furnished various impurity energy levels.

8.2.5 On Selecting Photocatalysis

Photocorrosion can be avoided to a decent extent by refraining from using systems which promote dissolution. Photoelectrodes with excessively high or low band gap value should be avoided as our focus is mainly on the visible part of the radiation. Abetting the promotion of any part of the Photo-redox reaction gives them both implicitly and stoichiometrically different products. An ionic character extent of 20 to 30% along with the requisite positioning of the valence band and conduction band edges may be considered ambient.

Due to the particularly favorable physiological properties such as high surface area, porous structure, small crystalline size and light-assisted sensitization possible, TiO_2 has rightly acquired almost one-third share of the total no of publications on the photocatalytic substances. In this, the anatase phase which happens to be the moderate phase is fabricated by the hydrolysis of Ti alkoxide (e.g., isopropoxide) followed by a controlled calcinations (Yan et al. 2013). However, it must be noted that TiO_2 is activated only under the UV light. This calls out for the incorporation of some doping and band gap engineering schemes to address the issue.

People have achieved appreciable results using metal doping strategies. In that case, the generated electron, which has a tendency to approach the metal which happens to be an efficient conductor, hence fulfills the purpose of generating as well as separating the carrier is efficient. Highly ordered studies carried on to dope TiO_2 with metals such as iron, molybdenum, vanadium, and ruthenium to name a few, have recorded high values of photocatalytic activity (Choi et al. 1994). Studies over Au loaded TiO_2 conclude that nano particles in the range of 30 nm resulted in increased hydrogen generation. Doping with lanthanide series ions in conjunction with the 3rd oxidation state exhibiting co-dopants such as Nd^{3+}/Eu^{3+}, Eu^{3+}/Ho^{3+}, or Nd^{3+}/Er^{3+} has positively affected the photocatalytic behavior (Reszczyńska et al. 2014).

Apart from the regular doping incorporations, CdS, due to its favorable band gap of about 2.5 eV is one of the most researched upon photocatalyst in its individual ambit too. One of the studies reports that Pt-load CdS, particularly in its hexagonal crystal structure, is more promising than its other structural variants (Matsumura et al. 1985). Different morphological modifications ranging from nanowires to nanosheets have been published to show high activity (Xu et al. 2013; Jang et al. 2007). In contrast to the virtually spacive choices like hollow nanospheres and nanorods, a solid nanosphere mitigates the recombination problem in a better way (Li et al. 2011). Modifications with noble metals especially Pt has been reported promising in the co-catalytic regime (Jang et al. 2007).

In a strikingly fruitful demonstration, calcinations over sufficiently high temperature and at an Ar employed atmosphere have taken care of recombination to an extent that 55.8 times higher photocatalytic activity was recorded, that too under visible radiation (Fan et al. 2011).

Doping the TiO_2 photoelectrode with CdS (a semiconductor) also is of significant importance here (Qin et al. 2015). With a visible light triggerable band gap of 2.5 eV as compared to 3.2 eV for that of TiO_2 can be made to generate the excitons. The electron can then be made to migrate at the conduction band of TiO_2 and the hole is retained at the valence band of the semiconductor (CdS, over here).

One of the most genuine solutions comes from the arena of organic dyes. Dyes which have a relatively lower band gap can be excited on to generate the exciton pair by the visible fraction of light. The electron can now be passed over to the conduction band of TiO_2 from that of the dye, which then can be used to reduce any acceptor molecule. In this process, the dye suffers its chromophoric photodegradation (Ajmal et al. 2014).

Band gap engineering in conjunction with doping too has some solutions to offer. By administering impurities, we can control the valence and conduction band's energy level in pristine TiO_2 and use them to aid the photocatalytic process. The transition metals if used to dope onto the TiO_2 surface can lower the conduction band (Ola and Maroto-Valer 2015). On the other hand, non-metals can raise the energy level of the valence band. The non-metal doped catalysts (such as the ones with N, C, B) create acceptor impurity levels thereby drastically reducing the band gap values (Wang et al. 2012). Here, halides have also been reported to enhance the activity (Wang et al. 2012).

The collective oscillation of conducting electrons in the metal nanostructures which forms the basis for the phenomenon of plasmonic resonance (Ola and Maroto-Valer 2015) can be used to sensitize the TiO_2 nanoparticles. As an example, plasmonic resonance can be seen in the case of Pt loaded TiO_2 (Zhao 2014). Gold, silver, and copper nanoparticles are known to exhibit the phenomenon of plasmonic resonance in the visible range. These can then serve as plasmonic nanoantennae (Liu et al. 2011) to capture solar energy, wherein the metal incorporations behave as electronic sinks enhancing the interfacial charge transfer.

Another important innovation in this regard is the metal decorated core shell incorporation which is able to harness the goodness of both the worlds viz of surface plasmonics as well as that of exciton trapping. Here, any metal exhibiting plasmonic resonance in the visible range is made the core (5–150 nm) over which a shell of metal (e.g., Pt) or metal oxide (e.g., RuO_2) is furnished. Finally, this arrangement is subjected to metal or metal oxide decorations. Given its morphological attributes, this arrangement is able to migrate the charge carriers quite satisfactorily. While in case of the metal, the hole remains at the core and the electrons are guided toward the periphery, the opposite happens for the hole in case of a metal oxide scenario, i.e., the holes are directed toward the periphery. However, this technique is rendered rather ineffective as the metal core is found to shorten the generated exciton pairs resulting in recombinations. This issue can be dealt appreciably with by introducing an interfacial layer of silica in between the core and the shell layers.

Also have been tried are the transition metals which are known to have shown variable oxidation states (e.g., Fe, Cu, V). The idea here is to choose a metal exhibiting variable oxidation state and dope the TiO_2 with oxide of that metal. The dopant creates additional impurity levels and hence effectively lowers the band gap of the region up to the visible ambit. The dopant also acts here as the electron trapping center, thus also helping the system with the successful migration of the photogenerated charge carriers.

The roles of corner shared octahedral units of metal cations networks have also garnered appreciable publication. The prominent examples being those of perovskites are $LaTiO_{3.5}$, pyrochlores, and the weberites (Mkhalid 2015). The only requirement over here is the unhindered presence of the metal cations. As an example, Gd_3TiO_7 has two metal cations namely GdO_6 and Tantalate, TaO_6, and this compound, due to the reasons mentioned, is inactive for the water splitting reactions. On the other hand, compound like $LaTaO_7$ shows reasonable photocatalytic activity (Cheen et al. 2010).

Publications also underline the role of the dipole moment shaping up due to the distortion in the structural symmetry of the substance. This actually gives rise to local internal fields, thereby creating dipoles. These distorted units contribute to the excitons upon their photoexcitation. As an example, distorted $BaGa_2O_4$ shows a net dipole moment and is photocatalytically active.

8.3 Electrochemistry-Based Catalysis

So far the systems being discussed were having light-assisted carrier generation activities. However, these systems come up with their innate requirement of light which makes their candidature unfit for night applications. Hence to complement the former catalytic options, it is very much required that the water splitting is made to be carried out using simple electrochemistry-based methods too.

Electrocatalysts are those chemical entities which facilitate the electrochemical reactions. They either can be used as a modification of the surfaces or can be employed as the electrodes for the reaction. They affect the rate of the reaction by adsorbing the reactant on their surfaces and thereby facilitate the formation of intermediates to render an easy charge transfer between the reactant and the electrode interface.

8.3.1 Oxygen Evolution Reactions

Electrocatalysis for the OER reactions has an important role to undertake which seems to be the reason for the evolution of reports citing numerous electrocatalysts which have enhanced the yield of oxygen. The section at hand provides some recent developmental insights into the material related advances in the field.

On one hand, the metal oxides are at the cornerstone of the basic mechanism of the OER reaction, while on the other hand, the non-oxide incorporations also show a keen prospect in increasing the overall water splitting.

Although the standard reduction potential of 1.23 V is required to achieve water splitting, yet a higher potential is required to overcome the kinetically generated barriers (say 2 V).

8.3.1.1 Minutes of the OER

As has been stated earlier, in spite of the thermodynamic requirement of the SRP of 1.23 V, the kinetic reasons call for an additional potential value. This is referred to as the overpotential.

Apart from the exception of IrO_2 (Qin et al. 2015; Liu et al. 2011), OER involves certain common steps of adsorption of O and OH over the catalytic surface.

Prima facie, we can write,

$$OH^- + CS \rightarrow OH^* + e^- \tag{8.3}$$

$$OH^* + OH^- \rightarrow O^* + H_2O + e^- \tag{8.4}$$

Here, CS refers to the catalytic site and the symbol * represents the adsorbed specie over the catalytic surface.

O_2 may hence be furnished by the following two possible routes:

The first method simply requires the coupling of two adsorbed oxygen molecules viz.

$$O^* + O^* \rightarrow O_2 \tag{8.5}$$

The second route involves the formation of the intermediate OOH* by the reaction of O* and OH^- followed by its combination with another OH^- to finally yield O_2.

$$O^* + OH^- \rightarrow OOH^* + e^- \tag{8.6}$$

$$OOH^* + OH^- \rightarrow O_2 + H_2O + e^- \tag{8.7}$$

To begin by exploring the oxide electrocatalysts for the OER, the oxides of Ruthenium and Iridium which are also known as the precious metal oxides have been reported to be rutile in structure. Both of these are regarded as one of the finest catalysts for OER in both acidic as well as basic solutions. It should be noted over here that Ru is less stable than Ir in the oxide state and the subsequent electrode will be damaged up under high anodic potential (Ismail and Bahnemann 2014). Schematic diagram of OER is shown in Fig 8.2.

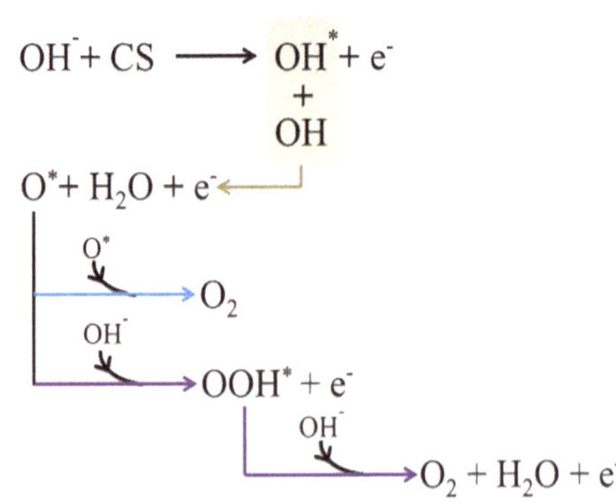

Fig. 8.2 Schematic diagram of OER

8.3.1.2 Perovskite Systems

Another widely explored domain of study is the perovskite family, having ABO_3 as the general formula with A being an alkaline element and B, a transition metal. Here, B occupies the center of the octahedron with a corner-shared arrangement. Loosely speaking the oxidation state values over here being +3 and +4 for B. A systematic study on a $La_{1-x}SrxFe_{1-y}Co_yO_3$ proposed that the OER activity would increase by increasing the x and y values (Matsumoto et al. 1980). The reason being given as the higher oxidation state of the Co ion coupled with the favorable distribution of the d character. In another staggering study over the systems at hand, a general trend of Ni > Co > Fe > Mn > Cr of the OER was proposed for varied transition metal candidates (Bockris and Otagawa 1984).

Promising candidature comes from the family of the layered-structured materials. Due to their structural attributes, the members of this family generally have an appreciable value of conductivity and have since a long time been finding roles in the energy storage arena. What is really peculiar to note here is, these compounds too have been noted to follow the same trend as that of the perovskite-based systems when tested for the OER activity for the case of the transition metal hydr (oxy)oxides (Bockris and Otagawa 1984).

Electrocatalysts which are incorporated with transition metals also have shown a reasonable ambit of OER activity. This behavior seems to exist in most of the scenarios comprising of different systems (Liu et al. 2015).

Switching over to some non-oxide electrocatalytic options the most noticeable candidates are the chalcogens, wherein the majority of the turf occupied by nickel- and cobalt-containing compounds. The additional qualities such as good conductivity and less deterioration against acid- and alkali-based environments are attributes of further advantage (McKendry et al. 2016; Yang et al. 2016). The coordination scheme exhibited over here is octahedral with some exceptions being tetrahedrally coordinated such as T_9S_8, Ni_3C_2 and T_3S_4 (Liu et al. 2014; Swesi et al. 2016). Metal di-chalcogenides such as $CoSe_2$ nanobelts have also been attended to (Liu et al. 2014).

Sufficient research work has also been undertaken by the incorporation of phosphorous- and nitrogen-based compounds of metals such as Co_4N, Ni_3N, NiP, CoP with their peculiarity being an appreciable degree of inertness in both acidic as well basic types of environments thus coming particularly handy for water splitting applications (Safizadeh et al. 2015; Xiao et al. 2015).

The usage of metals in the domain of electrocatalysis naturally gives invitation to various cost-based constraints the answers to which are expected to come from the non-metal arena. Carbon-based materials such as graphene, nanotubes, nanowires are subjected to doping by O, N, and also S for the sole reason of good electro-conductivity as well as their relatively inert modus operandi in the case of both acidic as well as alkaline working environments (Ullman et al. 2016).

Thus by the means of the aforementioned examples, the role of electrocatalysis is motivated. If used in conjunction with its light-assisted counterpart, it has a lot of promise to offer for alternative energy generation.

Catalyst	Electrolytes	Overpotential (mV)	Tafel slope (mV/dec)	Ref.
RuO_2	0.5 M H_2SO_4	240 @ 1 mA/cm^2	39	Mattos-Costa et al. (1998)
IrO_2	0.5 M H_2SO_4	290 @ 1 mA/cm^2	61	Audichon et al. (2016)
$La_{0.4}Sr_{0.6}FeO_3$	1 M KOH	622 @ 40 mA/cm^2	58	Matsumoto et al. (1980)
$LaFeO_3$	0.1 M KOH	400 @ 40 mA/cm^2	77	Zhu et al. (2016)
$MnFe_2O_4$	0.1 M KOH	470 @ 40 mA/cm^2	114	Li et al. (2015)
$CoCr_2O^4$/CNT	1 M KOH	326 @ 40 mA/cm^2	51	Al-Mamun et al. (2016)
NiCo-LDH	1 M KOH	300 @ 40 mA/cm^2	40	Song and Hu (2014)
FeNi-rGO LDH	1 M KOH	210 @ 40 mA/cm^2	40	Long et al. (2014)

8.4 Minutes of the HER

To achieve good electrocatalytic structures and to better understand the rate of the water splitting reaction, a fair appreciation of HER process becomes important. Like its counterpart OER, the HER also requires an overpotential to take care of the kinetic considerations. Its mechanism is highly pH-dependent. Let * denotes the adsorbed molecule on the catalytic site (CS), and the basic scheme of the reaction follows certain distinct steps.

The mechanism is initiated using the Volmer reaction which results in an adsorbed hydrogen due to the combination of a proton and an electron at the catalytic site. We thus have

$$CS + H^+ + e^- \rightarrow H^* \text{ (Volmer Reaction)} \quad (8.8)$$

Hydrogen gas may then be achieved by the following two routes:

$$H^* + H^+ + e^- \rightarrow H_2 + CS \text{ (Heyrovsky Reaction)} \quad (8.9)$$

Herein, the adsorbed hydrogen, in conjunction with a proton and an electron yields a hydrogen molecule.

The second method involves the coupling of two adsorbed hydrogen atoms to give a hydrogen molecule. This is known as the Tafel reaction:

$$H^* + H^* \rightarrow H_2 + CS \text{ (Tafel Reaction)} \quad (8.10)$$

It is important to note that above reaction schemes hold only for acidic environments.

As HER is a highly pH-dependent reaction, its mechanism gets altered in the following manner due to basic conditions. Owing to the diminished concentration of protons, H_2O in place of H^+ now gets coupled with an electron to yield the hydrogen atom adsorbed on the catalyst. We have

$$H_2O + e^- + CS \rightarrow H^+ + OH^- \text{ (Volmer Reaction)} \tag{8.11}$$

This adsorbed hydrogen, H*, now combines with water molecules to give molecular hydrogen and an electron on the catalytic site:

$$H^* + H_2O + e^- \rightarrow H_2 + OH^- + CS \text{ (Heyrovsky Reaction)} \tag{8.12}$$

The final reaction is the same as that in case of the acidic medium, wherein

$$H^* + H^* \rightarrow H_2 + CS \text{ (Tafel Reaction)} \tag{8.13}$$

Having gone through the minutes of HER reaction, let us now see different examples of various electrocatalysts used.

8.4.1 HER Electrocatalysts

8.4.1.1 Noble Metals

Noble metals such as Pt, Ir, Ru, Pd have been extensively studied (Li et al. 2016). Apart from the cost constraints, Pt has been reported to be one of the finest electrocatalysts for both acidic as well as basic environments. Increasing the number of active catalytic sites by achieving geometries in the nanorange seems to be a promising option (Xu and Zhang 2014; Guo and Wang 2011; Zhang and Li 2012). A nanocube with Pt shell on a Pd core over a graphene substrate has been reported by Bai et al. (2014). A current density of slightly over 790 mA cm^{-2} was reported at a potential of −300 mV. Achieving the catalytic structures through porous nanostructures is also done (Xu and Zhang 2014). The issues over here being easy dissolution and the inability in furnishing multi-metallic nanostructures using this technique (Zhang and Li 2012).

8.4.1.2 Alloys

Due to their industrially favorable advantages such resistance to corrosion and low cost, nickel based alloys have outperformed the ones using other elements (Safizadeh et al. 2015). Binary alloys using Ni over suitable mild steel substrates have been studied to conclude that their catalytic activity follows the order NiMo > NiZn > NiCo > NiW > NiFe > NiCr (Bai et al. 2014). Of these, the behavior of NiMo is particularly noteworthy because of its high stability as well as activity in its binary or ternary form (Safizadeh et al. 2015). In another recent study, a technique to synthesize NiMo nanopowders has been reported. The peculiarity over here being their unsupported character which when dissolved in suitable solvents could then be incorporated in any desired substrate (McKone et al. 2013). The subsequent Ti foil substrated NiMo alloy outperformed the others in the HER.

Apart from Ni, Fe-based alloys have also been reported to show good results particularly due to their hardness and appreciable resistance to corrosion. Crystalline alloys such as $Fe_{90}Ce_{10}$, $Fe_{90}Sm_{10}$, $Fe_{90}Y_{10}$, and $Fe_{90}M$ (M-Mischmetal) have been prepared with the mischmetal alloy showing the highest catalytic activity (Rosalbino et al. 2005). In a recent study, doping of CV pretreated Fe alloys with B and Si has reported an enhanced HER activity (Muller et al. 2014).

8.4.1.3 Carbides of Transition Metals

Due to their fair catalytic abilities, carbides, especially of tungsten and molybdenum, have been widely investigated (Wirth et al. 2012). An electrocatalytic activity comparison of nine transition metal carbides by Regmi et al. concluded that for HER, the activity of Mo_2C was contrastingly higher than the others (Regmi et al. 2015). They were followed subsequently by the carbides of tungsten and vanadium. Bulk metal carbides can also directly be used but at the cost of the decreased electrocatalytic activity. To increase the number of active sites, nano-ranged Mo_xC_y has been used. Nanoparticles of Mo_xC_y in various morphologies (nanorods, nanospheres, etc.) have been prepared in a recent endeavor by carburization of molybdenum materials contained in organic–inorganic amines (Wan et al. 2013; Ma et al. 2014).

8.4.1.4 Phosphides of Transition Metals

Transition metal phosphides, due to their easy availability in the earth crust, are also attracting research (Shia and Zhang 2016). Ever since their first study by Rodriguez et al. in 2005, they have been widely studied for both HER and OER under a variety of chemical environments. Recently, from a range of chemical phosphides species studied, those of Ni and Co have been reported for the most considerable candidatures. Electrodes made of transition metal phosphides with different organo-phosphine reagents were furnished to study their HER properties affirming almost all phosphides exhibited considerable activity with Ni_2P being most noteworthy (Read et al. 2016). The first and most common synthesis route was reported by Lewis et al. which employed tri-*n*-octylphosphene for phosphorous which in its liquid state, at a temperature of about 300 C, reacted with suitable precursors to render the respective phosphide of Ni and Co (Popczun et al. 2013; Popczun et al. 2014). The issue with the aforementioned method being the toxicity of either the raw material or the intermediate and also the expense incurred.

8.5 Conclusion

Hydrogen is a way cleaner, greener, and efficient fuel choice. Today, when the energy needs are constantly ramping up and the pollution levels are at an all time high, the day is not far when the conventional fuels would cease to offer an economical choice. Even a conservative estimate would show that the present trajectory of pumping in greenhouse gasses into the atmosphere is going to offer a very serious share of repercussions. Before a change of irreversible stature starts to get rampant, one has to arrive at a naturally congenial alternate source of energy which at the same time will not compromise with the world's developmental endeavors. Water-split hydrogen seems to offer the most reasonable and abundant choice for this problem. The present-day scenario, wherein around a measly 4% of the total hydrogen produced comes from the electrolysis, calls out for innovations to make the process feasible from the industrial point of view.

This chapter tried encompassing the recent developments in this sector both in the photocatalytic as well as the electrocatalytic domain. It was made implicit that catalysts would have to function as the sustainers of the economy. It was shown how UV range activated catalyst can be doped with a whole plethora of options to lower down its band gap upto an extent that it could be triggered by the visible light itself. The recombination needs to be taken care of by making ambient arrangements for exciton migration. Various schemes of catalytic activity enhancement were presented along with the principles behind them.

Gone are the days when nature could be assumed to be a commodity. The focus of research should be on building up innovations which could really help us in decreasing our reliability over nature. It is the time when one has to embark on a journey of decarbonisation by putting his best foot forward.

Acknowledgements D. K. Tiwari wishes to thank Dr. K. Rajeev Kumar and Dr. K. N. Madhusoodanan, Department of Instrumentation, Cochin University of Science and Technology for their support. A. Ray (IF140920) and S. Das (IFA13-PH-71) wish to thank INSPIRE, Department of Science and Technology (DST), Govt. of India for the financial support.

References

Ajmal A, Majeed I, Malik RN, Idriss H, Nadeem MA (2014) Rsc Adv 4:37003

Al-Mamun M, Su X, Zhang H, Yin H, Liu P, Yang H, Wang D, Tang Z, Wang Y, Zhao H (2016) Small 12:2866–2871

Audichon T, Napporn TW, Canaff C, Morais C, Comminges C, Kokoh KB (2016) J Phys Chem C 120:2562–2573

Bai S, Wang C, Deng M, Gong M, Bai Y, Jiang J, Xiong Y (2014) Angew Chem Int Ed 53:12120–12124

Bockris JOM, Otagawa T (1984) J Electrochem Soc 131:290

Cheen X, Shen S, Guo L, Mao SS (2010) Chem Rev 110:6503

Choi W, Termin A, Hoffmann MR (1994) J Phys Chem 98:13669

Chowdhury P, Malekshoar G, Ray AK (2017) Inorganics 5:34

Fan Y, Chen G, Li D, Li F, Luo Y, Meng Q (2011) Mater Res Bull 46:2338
Fujishima A, Honda K (1972) Nature 238:37
Guo S, Wang E (2011) Nano Today 6:240–264
Ismail AA, Bahnemann DW (2014) Sol Energy Mater Sol Cells 128:85
Jang JS, Joshi UA, Lee JS (2007) J Phys Chem C 111:13280
Jeon NJ, Noh JH, Kim YC, Wang WS, Ryu S, Seok SI (2014) Nat Mater 13:897
Lee R (2011) Science 333:569
Li C, Yuan J, Han B, Shangguan W (2011) Int J Hydrogen Energy 36:4271
Li M, Xiong Y, Liu X, Bo X, Zhang Y, Han C, Guo L (2015) Nanoscale 7:8920–8930
Li X, Hao X, Abudula A, Guan G (2016) J Mater Chem A 4:11973–12000
Liu Z, Hou W, Pavaskar P, Aykol M, Cronin SB (2011) Nano Lett 11(3):1111
Liu Y, Cheng H, Lyu M, Fan S, Liu Q, Zhang W, Zhi Y, Wang C, Xiao C, Wei S, Ye B, Xie Y (2014) J Am Chem Soc 136:15670
Liu Y, Wang H, Lin D, Liu C, Hsu PC, Liu W, Chen W, Cui Y (2015) Energy Environ Sci 8:1719
Long X, Li J, Xiao S, Yan K, Wang Z, Chen H, Yang S (2014) Angew Chem Int Ed 53:7584–7588
Ma Y, Guan G, Phanthong P, Hao X, Huang W, Tsutsumi A, Kusakabe K, Abudula A (2014) J Phys Chem C 118:9485–9496
Markham MC (1955) J Chem Educ 32:540–543
Matsumoto Y, Yamada S, Nishida T, Sato E (1980) J Electrochem Soc 127:2360
Matsumura M, Furukawa S, Saho Y, Tsubomura H (1985) J Phys Chem 89:1327
Mattos-Costa FI, de Lima-Neto P, Machado SAS, Avaca LA (1998) Electrochim Acta 44:1515–1523
McKendry IG, Thenuwara AC, Sun J, Peng H, Perdew JP, Strongin DR, Zdilla MJ (2016) ACS Catal 6:7393
McKone JR, Sadtler BF, Werlang CA, Lewis NS, Gray HB (2013) ACS Catal. 3:166–169
Mkhalid IA (2015) J Alloy Compound 613:298
Muller CI, Rauscher T, Schmidt A, Schubert T, Weißgärber T, Kieback B, Rontzsch L (2014) Int J Hydrogen Energy 39:8926–8937
Ola O, Maroto-Valer MM (2015) Appl Catal A 502:114
Popczun EJ, McKone JR, Read CG, Biacchi AJ, Wiltrout AM, Lewis NS, Schaak RE (2013) J Am Chem Soc 135:9267–9270
Popczun EJ, Read CG, Roske CW, Lewis NS, Schaak RE (2014) Angew Chem Int Ed 53:5427–5430
Qin N, Liu Y, Wu W, Shen L, Chen X, Li Z, Wu L (2015) Langmuir 31(3):1203
Read CG, Callejas JF, Holder CF, Schaak RE (2016) ACS Appl Mater Interfaces 8:12798–12803
Regmi YN, Waetzig GR, Duffee KD, Schmuecker SM, Thode JM, Leonard BM, Mater J (2015) Chem A 3:10085–10091
Reszczyńska J, Grzyb T, Sobczak JW, Lisowski W, Gazda M, Ohtani B, Zaleska A (2014) Appl Surf Sci **307**:333
Rosalbino F, Maccio D, Angelini E, Saccone A, Delfino S (2005) J. Alloys Compd 403:275–282
Safizadeh F, Ghali E, Houlachi G (2015) Int J Hydrogen Energy 40:256
Shia Y, Zhang B (2016) Chem Soc Rev 45:1529–1541
Shockley W, Queisser HJ (1961) J Appl Phys 32:510
Song F, Hu X (2014) Nat Commun 5:4477
Swesi AT, Masud J, Nath M (2016) Energy Environ Sci 9:1771
Ullman AM, Brodsky CN, Li N, Zheng S-L, Nocera DG (2016) J Am Chem Soc 138:4229
Walter MG, Warren EL, McKone JR, Boettcher SW, Mi QX, Santori EA, Lewis NS (2010) Chem Rev 110:6446
Wan C, Knight NA, Leonard BM (2013) Chem Commun 49:10409–10411
Wang DH, Jia L, Wu XL, Lu LQ, Xu AW (2012) Nanoscale 4:576
Wirth S, Harnisch F, Weinmann M, Schroder U (2012) Appl Catal B 126:225–230
Xiao P, Chen W, Wang X (2015) Adv Energy Mater 5:1500985
Xu Y, Zhang B (2014) Chem Soc Rev 43:2439–2450

Xu Y, Zhao W, Xu R, Shi Y, Zhang B (2013) Chem Commun 49:9803
Xu Y, Kraftacd M, Xu R (2016) Chem Soc Rev 45:3039
Yan H, Wang X, Yao M, Yao X (2013) Prog Nat Sci Mater Int 23(4):402
Yang J, Zhu G, Liu Y, Xia J, Ji Z, Shen X, Wu S (2016) Adv Funct Mater 26:4712
Yin W, Bai L, Zhu Y, Zhong S, Zhau L, Li Z, Bai S (2016) ACS Appl Mater Interfaces 8 (35):23133
Zhang J, Li CM (2012) Chem Soc Rev 41:7016–7031
Zhang Z, Yates JT Jr (2012) Chem Rev 112:5520
Zhao Z-Y (2014) J Phys Chem C 118:24591
Y. Zheng, L. Lin, B. Wang, X. Wang, *Angew.* (2015) *Chem. Int. Ed.*, *54*, 12868
Zhu Y, Zhou W, Yu J, Chen Y, Liu M, Shao Z (2016) Chem Mater 28:1691–1697

Chapter 9
New Age of Wastewater Treatment Employing Bio-electrochemical Systems

M. M. Ghangrekar and Pritha Chatterjee

Abstract Bio-electrochemical systems (BESs) as an energy-efficient wastewater treatment technology have attracted increasing interest in the past decade. A BES is a bio-electrochemical reactor capable for conversion of chemical energy present in the chemical bonds of organic compounds directly to other usable forms of chemical energy or electrical energy by using catalytic reactions initiated by micro-organisms in anaerobic conditions. Energy crisis in recent time has attracted researcher's interest in BESs as an option to produce electric power or other forms of energy from biomass without producing any carbon emission to the environment. BESs can also be utilized at wastewater treatment facilities to treat different types of organic matter present in the wastewater and recover valuable products, thus ensuring economic sustainability. This chapter comprehensively reviews BES with emphasis on their applications for wastewater treatment. It is anticipated that insights offered in this chapter would facilitate for the use of BES as an energy-efficient wastewater treatment option coupled with direct electricity or recovery of resources.

Keywords Bio-energy · Bio-electrochemical systems · Electricity generation
Wastewater treatment

9.1 Introduction

Partially treated or untreated wastewater generated from human settlements as well industrial or commercial establishments results into major environmental contamination. A recent report of CPCB (CPCB 2013) states that over 100 million m^3 of untreated or partially treated wastewater is discharged per day into the Ganga River

M. M. Ghangrekar (✉) · P. Chatterjee
Department of Civil Engineering, Indian Institute of Technology, Kharagpur 721302, India
e-mail: ghangrekar@civil.iitkgp.ernet.in

S. Bhattacharya et al. (eds.), *Water Remediation*, Energy, Environment, and Sustainability, https://doi.org/10.1007/978-981-10-7551-3_9

and its tributaries, which (the Ganga and its tributaries) functions as one of the major water resource of India. The main source of this wastewater is sewage generated from human settlements. The wastewater can be discharged into the open water sources like rivers and streams and can also seep into soil through faulty joints in sewers or while flowing along natural drains. This situation causes pollution in surface water bodies, deteriorates groundwater quality and finally leading to harmful health conditions for the society. The clean water also gets contaminated by these pollutants causing a severe decrement in clean water availability, causing a difficult socio-economic problem.

The importance of wastewater management is being recognized only in the last two decades in developing countries, and wastewater collection systems and treatment plants are being built to counteract the problem. Wastewater treatment plants (WWTPs) are dedicated to treat the wastewater discharged by human household, commercial and industrial sectors for considerably reducing the level of pollutants before discharging the treated wastewater to water bodies (Chen and Chen 2013). The main objectives of sewage treatment are to improve public health, reduce the oxygen demand of treated water before being discharged into water bodies, decrease eutrophication of receiving water bodies, remove pathogen and sometimes remove other compounds depending on the intended place of discharge or reuse (Chung et al. 2015).

The different wastewater treatment technologies adopted presently can be classified into natural system (oxidation pond—OP, waste stabilization pond—WSP, aerated lagoon—AL), conventional system (activated sludge process—ASP, extended aeration—EA, upflow anaerobic sludge blanket (UASB) reactor) and advanced technologies (sequencing batch reactor—SBR, biofar, moving bed bioreactor—MBBR, submerged aerobic filter—SAF, kernel technology—KT) (Chatterjee et al. 2016). Conventional sewage treatment processes demand high land requirement or capital investment, higher maintenance and operational cost and enormous energy requirements, which makes them unsuitable for use in developing countries (Sato et al. 2006).

Energy-efficient low-cost wastewater treatment systems are the best choice for such countries. Anaerobic treatment systems excel in such respect. The classical anaerobic processes commonly used are septic tanks or Imhoff tanks, anaerobic pond, anaerobic filter, expanded bed reactor and UASB reactor. Organic matter removal efficiency of septic tanks or Imhoff tanks is very low, and even after providing higher retention time, they are unable to achieve desirable efficiency, thus rendering the technology unattractive. Anaerobic ponds though can demonstrate reasonable organic matter removal efficiency; however, they occupy huge land area. Anaerobic filter faces operational problems due to blockage. Fluidized bed reactor requires additional energy for pumping to keep the media in expansion (Van Haandel and Lettinga 1994). Consequently, UASB reactor appears to be the most appropriate choice for economical sewage treatment, and it has been employed for treatment of sewage in many cities in India and abroad.

However, at many places, UASB reactor has failed to meet up to the expectations of the treatment efficiency, and either the performance is inferior due to high sulphate and suspended solids concentration in the raw sewage received or there are problems related to design and operation of these reactors. Aerobic wastewater treatment plants either need very large land area (pond systems) or enormous energy for aerating the wastewater (ASP). Hence, there is a need to have paradigm shift in thinking and rather than considering wastewater as waste, consider it as a resource and explore possibility to recover energy/other products from it while offering treatment to it. This approach will make wastewater treatment sustainable, and it will certainly contribute in solving water pollution problem. The research efforts in last two decades have brought new technological solution for treatment of the wastewater by exploring bio-electrochemical approach. This has resulted in development of bio-electrochemical system (BES), which is an upcoming technology that has the potential to address simultaneously the major problems of energy crisis and waste management as either energy is generated in the form of direct electricity or some resources are recovered from the waste or both while treating wastewater (Angenent et al. 2004).

9.2 Working Principle of BES

BESs are emerging technologies where micro-organisms are encouraged to grow on one or both electrode(s) to catalyse the oxidation reaction in case of a bio-anode and/or reduction reaction in case of a biocathode (Rabaey 2010). Typical BES consists of an anode, a cathode and a proton- or cation-exchange membrane separating the two electrode chambers (Rabaey and Rozendal 2010). The electrodes are immersed in an electrolyte which is mainly the fluid around the electrode having the reactants and/or products (it is generally an aqueous solution or wastewater, which is intended to be treated) (Fig. 9.1). An oxidation reaction takes place at the anode (e.g. oxidation of acetate or oxidation of other organic matter), whereas a reduction reaction starts at the cathode (e.g. O_2 reduction or H_2 evolution). Soluble organic matter such as monosaccharide, acetate derivatives, alcohols and others present in wastewater can be oxidized by electrochemically active bacteria (EAB) present in the anodic chamber, and electrons and protons are produced (Wang and Ren 2013). The electrons are then carried to the anode surface by redox carriers such as nicotinamide adenine dinucleotide (oxidized), NAD^+; flavin adenine dinucleotide (oxidized), FAD^+; flavin mononucleotide (oxidized), FMN^+ or direct transfer of electron from bacterial cell to the electrode through cytochrome-c, which helps in capturing the energy during bacterial respiration (Kim and Gadd 2008; Nelson et al. 2008). These electrons flow through an external circuit, producing electricity, and finally react at the cathode for oxygen reduction reaction to form water and other chemical products or with other chemical catholyte to produce value-added chemicals. Contaminants such as uranium, chlorinated solvents and perchlorate can be remediated by using the electrons present in the cathodic chamber (Aulenta et al. 2008; Butler et al. 2010; Gregory and Lovley 2005). This potential developed across these electrodes can also be utilized for desalination of water by using a cell

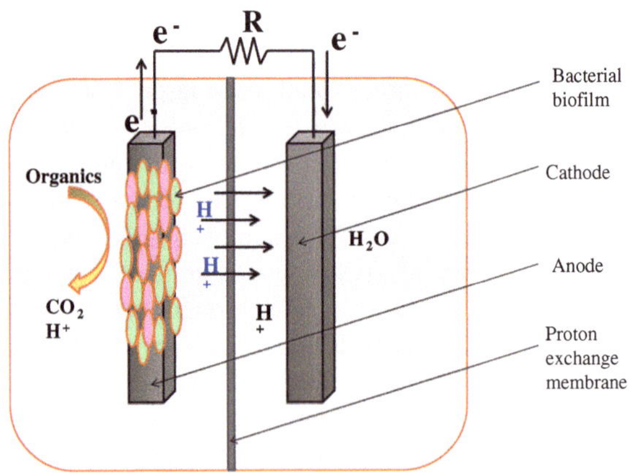

Fig. 9.1 Schematic of a bio-electrochemical system (BES)

configuration called as microbial desalination cell (Kim and Logan 2011; Luo et al. 2010).

9.3 Types of BES

BESs are classified as microbial fuel cell (MFC), microbial remediation cell (MRC), microbial chemical cell (MCC), sediment microbial fuel cell (SMFC), microbial electrolysis cell (MEC) and microbial desalination cell (MDC). The possible substrates that can be used, electron acceptors used and functions of each one of these are illustrated in Table 9.1.

9.3.1 Microbial Fuel Cell

Microbial fuel cells use microbial metabolism to convert the chemical energy present in organic matter to electrical energy. MFC can be effectively used for removal of organic matter from wastewater and recovering electricity as by-product. MFCs have been used as biochemical oxygen demand (BOD) sensor based on the principle of proportional correlation between electricity generation and organic matter concentration in the wastewater (Kim et al. 2003). MFCs can be used to provide long-term stable power to biomedical devices implanted in the human body, which typically have low power demand. Researchers have reported a mediated MFC for the detection and enumeration of bacteria (Kim and Gadd 2008). Such techniques have potential application for the rapid detection of micro-organisms in food, and in particular useful for detection of food contamination.

Table 9.1 Types of BES based upon the products formed at cathode

Types of BES	Substrate used for oxidation in anode	Electron acceptor at cathode	Function	References
Microbial fuel cell (MFC)	Domestic/industrial wastewater containing biodegradable organic matter	O_2, $K_3Fe(CN)_6$, HOCL, other oxidizing compounds	Electricity generation	(Jadhav et al. 2014; Logan 2008)
Microbial electrolysis cell (MEC)	Domestic/industrial wastewater containing biodegradable organic matter	H^+, CO_2, organics	H_2, H_2O_2, CH_4, NaOH, struvite synthesis	(Cheng et al. 2009; Cusick and Logan 2012; Rozendal et al. 2009)
Microbial desalination cell (MDC)	Domestic/industrial wastewater containing biodegradable organic matter	O_2, $K_3Fe(CN)_6$, HOCL, other oxidizing compounds	Electricity generation, desalination	(Kim and Logan 2011; Luo et al. 2010)
Microbial remediation cell (MRC)	Contaminated sediments with sulphate and nitrate for passive oxidation	O_2	Bioremediation of hydrocarbons	(Butler et al. 2010; Morris and Jin 2007)
Microbial solar cell (MSC)	Organic matter produced by photosynthetic organisms using sunlight	O_2, $K_3Fe(CN)_6$, HOCL, other oxidizing compounds	Electricity generation	(De Schamphelaire et al. 2010; Helder et al. 2010; Schamphelaire et al. 2008)
Microbial chemical cell (MCC)	Domestic/industrial wastewater containing biodegradable organic matter	CO_2, organics	Synthesis of organic compounds	(Nevin et al. 2008; Steinbusch et al. 2009)

9.3.2 *Microbial Desalination Cell*

MDC is a recent type of BES capable of offering simultaneous organic matter removal, desalination and energy production. MDC is formed by modifying the MFC by introducing few ion-exchange membranes (IEMs) between anode and cathode (Fig. 9.2). MDC is capable of desalinating sea water and saline wastewater partly, hence it can be used as a stand-alone system where low desalination efficiency is acceptable or it can be used as a pre-treatment option before the costly conventional technologies such as electro dialysis (ED), ion exchange, and reverse osmosis (RO). MDC typically has three chambers: anodic chamber, middle desalination chamber and cathodic chamber separated by anion-exchange membrane (AEM) and cation-exchange membrane (CEM) in different combinations.

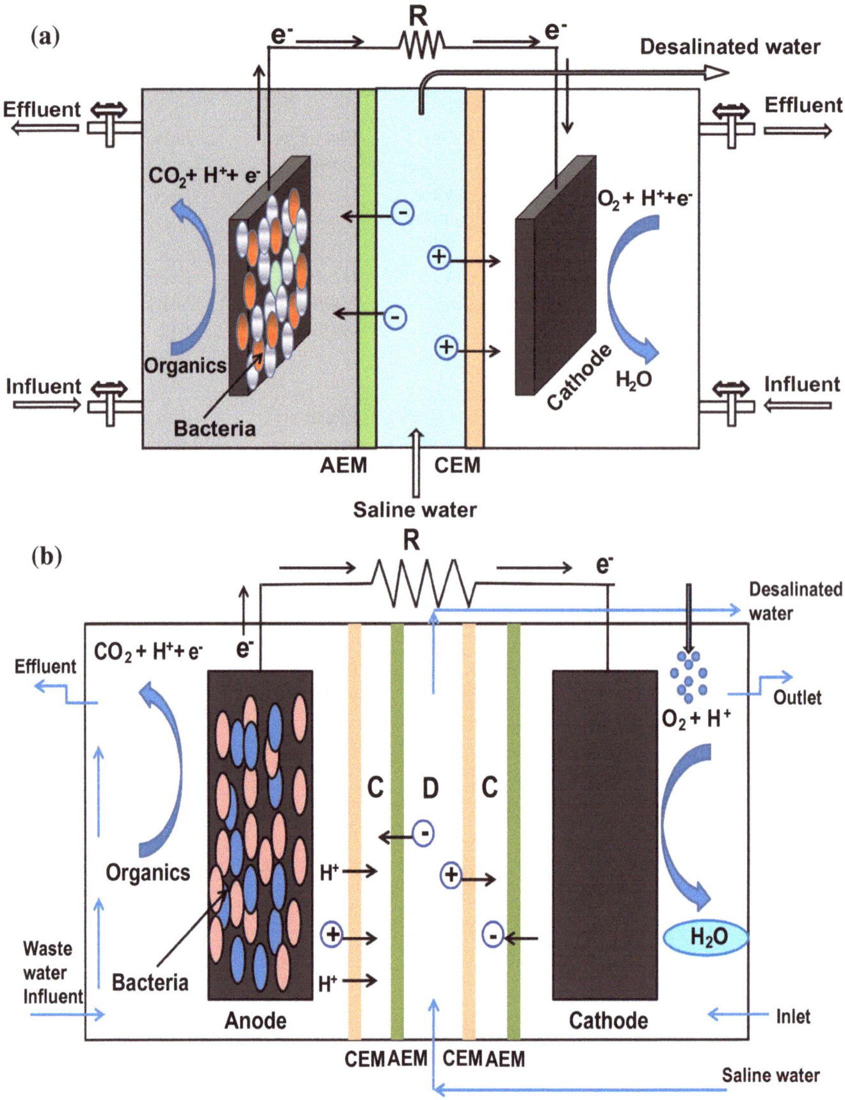

Fig. 9.2 Schematic of **a** three-chambered MDC and **b** five-chambered MDC

MDC having three chambers with AEM near anode exhibits better desalination performance as compared to placing CEM near anode showing higher columbic efficiency (CE) (Pradhan and Ghangrekar 2014b). However, the chemical oxygen demand (COD) removal was reported to be more when CEM is placed near anode. MDC works in the combined principle of MFC and ED. Pradhan and Ghangrekar (2014a) operated a five-chamber MDC with anode, cathode, one central desalination chamber and two concentrate chambers, located adjacent to desalination chamber, separated by ion-exchange membranes (Fig. 9.2b). Average removal efficiency for COD of 81 ± 2.1% and TDS of 58, 70 and 78% was reported with salt concentration, in the brackish water to be desalinated, of 8, 20 and 30 g/L, respectively, in the middle desalination chamber. The bio-electric potential developed between the anode and cathode in a MDC transports the ions present in the saline water towards the adjacent chambers from middle chamber through ion-selective membrane and desalinates the liquid waste. This low energy desalination technology is very similar to ED. External voltage is applied in ED; while in MDC, internal voltage developed, from oxidation of organic matter present in wastewater, helps in driving desalination.

9.3.3 Microbial Electrolysis Cell

A microbial electrolysis cell (MEC) is an emerging technology capable of electrochemically converting the soluble organic matter to storable energy in the form of hydrogen. The electrons supplied by bacteria in anodic chamber to anode are combined with the protons on cathode to generate hydrogen gas by addition of small power source. Theoretically, the cell voltage required for producing hydrogen at the cathode is 0.114 V, but it requires about 0.25 V due to occurrence of unavoidable overpotential. However, the requirement of supply of external voltage is still lower in comparison with the voltage required for water electrolysis (which is practically over 1.8 V); therefore, MEC can save energy consumption for hydrogen production (Logan 2008).

Since the invention of MECs in early 2005–06, rapid progress has been made by the researchers to develop this technology. Though the anode of a MEC is similar to that of MFC, the cathode is configured similar to water electrolysis process. For enhancing efficiency of a MEC, improvisation of the cathode is required by using transition metal and/or alloys as catalyst to increase its conductivity and to increase the efficiency of the cathodic reaction by improving catalytic activity and reducing charge transfer resistance. MEC is gaining popularity because microbes can be used as anode catalyst, it can be operated at ambient temperature and atmospheric pressure, and wastewater can be used as fuel. Biohydrogen production from MEC in this way is expected to play an important role in future economy.

MECs are a comparatively new technology for producing hydrogen gas from acetate and other fermentation by-products of electrohydrogenesis (Pant et al. 2012). Generally, in a MFC, current is produced by oxygen reduction reaction, in presence of oxygen at cathode; however in MEC, spontaneous current generation is not possible because cathode is anaerobic and in absence of oxygen. Thus, a small external voltage (typically 0.5–0.7 V) is required to be applied to the circuit, to support the production of hydrogen at the cathode through the reduction of protons. The anodic oxidation reaction is the same as in the microbial generation of electricity in MFC, while the cathodic reaction proceeds in absence of oxygen. Compared to dark fermentation reactor producing hydrogen from wastes, MEC has a higher hydrogen recovery and a wider substrate diversity (Sun et al. 2009b). Very recently, the performance of a pilot-scale continuous flow MEC (1000 L) fed with winery wastewater was reported to produce hydrogen in the range of 2.9 ± 0.2 mL L^{-1} d^{-1}, respectively.

9.3.4 Sediment Microbial Fuel Cell

SMFCs or benthic microbial fuel cells (BMFCs) produce electricity by placing an anode into organic-rich sediments and a cathode in the overlying water. Degradation of organic compounds present in the sediment and wastewater occurs due to presence of electrogenic bacteria, producing electrons and protons. Electrons cause reduction of the anode and are transferred to the cathode through an external circuit. The protons start moving from sediment to the cathode and combine with oxygen and electrons at cathode to produce water. In situ bioremediation of contaminated organic-rich sediment and overlaying water is performed by SMFC for the removal of hydrocarbons. This way the power generated from SMFCs can be effectively used to power remote sensors through power management system. Performance of SMFCs was also evaluated to understand the effect of the presence of cellulose in an aquaculture pond sediment by Sajana et al. (2014). Even in the presence of cellulose up to 2% in sediments, these SMFCs demonstrated efficient cellulose degradation while maintaining the oxidized sediment top layer favourable for practising aquaculture.

MRC is SMFC used for the treatment of contaminated sediments with sulphate and nitrate for passive oxidation along with energy recovery. On the cathode, sulphates and nitrates act as terminal electron acceptors (TEA) and their availability limits the rate of bioremediation of hydrocarbons. This MRC system can be deployed in a contaminated area for in situ remediation with anaerobic sediments and anaerobic overlying water, which would be applicable to locations such as shallow surface water impoundment, e.g. lagoon, estuary, and marsh and sensitive areas where dredging becomes difficult to facilitate removal of excessive hydrocarbon.

9.4 Practical Applications of BES for Wastewater Treatment

9.4.1 Sewage Treatment

MFCs are potential solution for wastewater treatment and electricity recovery using a variety of wastewaters including both domestic and industrial wastewaters (Aelterman et al. 2006; Ghangrekar and Shinde 2007). Ghangrekar and Shinde

(2008) evaluated long-term performance of a mediator-less and membrane-less MFC (ML-MFC) for treatment of sewage and achieved a maximum COD removal efficiency of 82.7% while producing a maximum current of 0.17 mA. A maximum power density with respect to cathodic surface is of 6.73 and 13.65 mW/m^3 with respect to anodic volume, and a maximum current density with respect to cathodic surface area of 70.74 mA/m^2 was reported. Treatment of domestic wastewater was performed under two operating temperatures (23 ± 3 and 30 ± 1 °C), and two different flow modes (i.e. fed-batch and continuous) using single-chamber air-cathode MFC were examined by Ahn and Logan (2010).

9.4.2 Nitrogen Removal

In a BES, nitrogen removal may include assimilatory nitrogen uptake, dissimilatory nitrate reduction and physicochemical factors such as ammonia volatilization at the cathode, which increases in proportion to current generation. A single-chamber air-cathode MFC fed with swine wastewater after 100 h of operation removed ammonia by 83 ± 4% (from 198 ± 1 to 34 ± 0 mg NH_4^+-N/L), whereas the soluble COD was reduced from 1240 ± 20 to 120 ± 30 mg/L (86 ± 6% removal) (Min et al. 2005). Nitrite and nitrate concentrations increased from 0.4 ± 0.1 to 2.9 ± 0.1 mg NO_2-N/L and 3.8 ± 1.2 to 7.5 ± 0.1 mg NO_3-N/L in the anolyte due to probable nitrification, likely as a result of oxygen diffusion through the cathode. Dual-chamber MFCs made of clay ware cylinder, where wall material of cylinder worked as separator, demonstrated effective ammonia and organic matter removal at lower COD:NH_4^+–N ratios (Jadhav and Ghangrekar 2016). With decreasing COD:NH_4^+ ratio, ammonia removal rate increased with maximum of 63% removal at a ratio of 1:1. Average COD removal efficiencies demonstrated by these MFCs were about 88%, demonstrating effective use of MFCs for treatment of wastewater containing both organic matter and high ammonia concentration. The ammonium-N removal slightly increased when microbes were exposed to only ammonium as a source of electron when organic matter was not supplemented. When operated with imposed potential on cathode and without aeration in the cathodic chamber, oxidation of ammonium ions at a faster rate observed in this MFC confirmed anaerobic oxidation.

9.4.3 Sulphur Recovery

Municipal sewage typically contains 20–500 mg/L sulphate. Many industries such as paper and pulp, food processing, animal husbandry and pharmaceutical produce the wastewater which usually contains much higher concentrations of sulphate, sulphite or other sulphur compounds (Lens et al. 1998; Zhao et al. 2008). Sulphate being chemically inert, non-volatile and non-toxic compound, these emissions are not a direct threat to the environment as sulphate. Present restrictions on sulphate emission in environmental legislation mainly aim to reduce the salt content of surface waters and/or to minimize acid condensation in sewers (Lens et al. 1998). Sulphate-reducing bacteria (SRB) reduce sulphate to sulphide under anaerobic conditions (Lee et al. 2014). Further studies are required to establish it as a suitable process for high sulphate-containing wastewater treatment (Dutta et al. 2009; Kelly and He 2014). In a study by Rabaey et al. (2006), a MFC with a hexacyanoferrate cathodic electrolyte was used to convert dissolved sulphide to elemental sulphur.

9.4.4 Industrial Wastewater Treatment

Wastewaters coming from different industrial operations contain high concentration of organic and inorganic substances as well as soluble and insoluble materials causing significant polluting phenomena. For example, in rice milling industry, the wastewater yield is about 1.0–1.2 L/kg of paddy, which was treated in a MFC fabricated using a clay ware ceramic pot, forming anodic chamber, and a COD removal efficiency of 96.5%, lignin removal efficiency of 84% and phenol removal efficiency of 76% have been reported (Behera et al. 2010). Maximum sustainable volumetric power of 2.3 W/m^3 was observed with 100 Ω external resistance. More effective treatment of rice mill wastewater and higher energy recovery was demonstrated by the cheaper clay ware pot MFC as compared to MFC incorporated with Nafion as proton-exchange membrane.

Food-processing wastewaters are characterized as non-toxic because they contain very less hazardous compounds, have moderate-to-high BOD compared to many other industrial wastewaters, and much of the organic matter in them consists of simple sugars and starch. Electricity production from food-processing wastewater was explored by Oh and Logan (2005) while treating cereal wastewater (diluted to 595 mg COD/L) in a two-chambered MFC, a maximum power of 81 $\pm$ 7 mW/m^2 (normalized to the anode surface area), current production of 25 $\pm$ 2 mA per litre of wastewater treated were reported, producing effluent with a final COD of less than 30 mg/L (95% removal). While using a single-chambered MFC and pre-fermented wastewater, the maximum power density of 371 $\pm$ 10 mW/m^2 (53.5 $\pm$ 1.4 mA per litre of wastewater) was reported.

Manufacture of starch-based products involves significant usage of water, consequently resulting in large quantities of wastewater. Generally, the COD concentrations of such wastewater range from 6000 to 10,000 mg/L, which can impose heavy load on the environment if discharged untreated. In a study by Lu et al. (2009), an air-cathode MFC could treat starch-processing wastewater containing 4852 mg/L of COD. Maximum voltage output and power density of 490.8 mV and 239.4 mW/m^2, respectively, (a current density of 893.3 mA/m^2) were reported with a maximum coulombic efficiency of 8.0%. Removal efficiencies of COD and ammonia-nitrogen amounted to 98.0 and 90.6%, respectively.

The performance of single-chamber membrane-free MFC for brewery wastewater treatment and electricity production was investigated by Wang et al. (2008). An attempt was made to provide solution for malt distillery wastewater using fungal pre-treatment by *Aspergillus awamori* followed by further treatment in MFC to achieve higher organic matter removal, which is a challenge at present with currently adopted technologies (Ray and Ghangrekar 2016). This combined process could successfully recover chitosan as 0.6–0.7 g/L of settled sludge (3.8% solids). This two-stage process demonstrated overall 99% COD removal as well as almost complete removal of SS.

9.4.5 Remediation of Aquaculture Wastewater in SMFC

Limited availability and higher cost of good quality water for practising aquaculture and the environmental impacts of used water discharges have led to the treatment and reuse of the aquaculture water contaminated with dissolved organics and ammonia rather than discharging it. The effectiveness of in situ SMFC for aquaculture water treatment along with generation of electricity was explored by Sajana et al. (2017). The SMFC demonstrated satisfactory COD, total Kjeldahl nitrogen (TKN) and total nitrogen (TN) removal efficiency of $79 \pm 1, 94 \pm 3$ and $62 \pm 4\%$, respectively. This SMFC offered successful in situ treatment to the aquaculture pond water. This SMFC system will have the potential to drastically reduce the cost of water treatment apart from savings in the pumping cost, which is inevitable when the aquaculture water is treated externally and reused. As expected, very low power was generated by the SMFCs that can be only used for powering the sensors; however, it demonstrated successful in situ remediation of aquaculture water.

9.4.6 Treatment of Animal Waste

For sustainable animal production, treatment and odour control of wastewater generated are important components in swine farming. A single-chamber MFC was used by Kim et al. (2008) to reduce 10 chemicals associated with odours by 99.76% (from 422 ± 23 μg/mL) and three volatile organic acids (acetate, butyrate and propionate) by greater than 99%. The MFC produced a maximum power of 228 mW/m^2 and removed 84% of the organic matter during 260 h of retention time.

9.5 Bottlenecks, Challenges and Future Scope

Two redox couples are required in MFC: first for coupling the reduction of an electron mediator to a microbial oxidative metabolism and the second for coupling oxidation of the electron mediator to the reduction of the electron acceptor on the cathode surface, where the electron acceptor is regenerated with atmospheric oxygen. The ideal performance of an MFC depends on the electrochemical reactions that occur between the organic substrate at a low potential such as glucose and the final electron acceptor with a high potential, such as oxygen (Rabaey and Verstraete 2005). The actual cell potential is always lower than its equilibrium potential because of irreversible losses, namely activation polarization losses on anode and cathode, concentration polarization in cathodic and anodic chambers and ohmic losses because of resistance to the flow of ions in the electrolyte and resistance to flow of electrons through the electrode. Activation polarization is caused because of an activation energy that must be overcome by the reacting species. Processes involving adsorption of reactant species, transfer of electrons across the double-layer cell membrane, desorption of product species and the physical nature of the electrode surface all contribute to the activation polarization (Du et al. 2007). In an MFC, power density that can be achieved depends on several factors and hence optimization of geometrical configuration of MFCs, electrogenic bacterial consortium selection and enrichment, substrate used, concentration of bacteria, electron shuttles, electrochemical properties of electrodes, electrode surface area, etc., are required.

Anode is a crucial component of the MFC set-up, both structurally and functionally. Apart from providing support for bacterial attachment, it acts as a sink for electrons from substrate metabolism. Poor performance of anode in MFC is still a major setback for its practical applications (Hindatu et al. 2017). Meagre microbial attachment or attachment of non-electrogenic bacteria on the electrode surface limits the effectiveness of electron transfer between the micro-organism and the anode, thus affecting the overall performance of MFC resulting in low power output, and hinders practical acceptance of MFC technology. Other contributing factors include high internal resistance, which is attributed to charge and mass transfer losses and high cost of the electrode materials. Anode material should favour enhanced microbial development/attachment, substrate metabolism and extracellular electron transfer and at the same time should not be expensive (Hindatu et al. 2017). Additionally, the material selected for anode should also exhibit good electrical conductivity with low resistance and large surface area. Furthermore, it should be chemically inert, should offer resistance to corrosion and degradation, and should possess suitable mechanical strength and toughness (Hindatu et al. 2017).

The base material for the cathode provides a conductive path to the external circuit. To function as a good electrode, a cathode material should have the properties, namely (a) good conductivity, (b) high chemical stability, (c) high mechanical strength and (d) low cost. Carbon, in paper or cloth form, is the most widely used material in air-cathode MFCs (Ahmed et al. 2012; Cheng et al. 2006; Martin et al. 2011; Shi et al. 2012; Sun et al. 2009a; Zhang et al. 2009).

For scaling up of MFCs, it is necessary to know the maximum possible current generation supported under optimum working conditions, which is governed by bio-electrochemical parameters for a given anodic chamber volume treating a particular wastewater. Some of the challenges associated with the scaling up of MFCs are discussed in the literature. These challenges include maintaining low internal resistance of MFC while increasing the activity levels of electrochemically active biomass and maintaining it for longer duration of operation, optimization of reactor design, selection of low-cost electrode materials and catalysts that can support higher and stable current densities and developing newer innovative ways of separating anode from cathode in a cost-efficient way.

9.6 Conclusion

The building of a sustainable society will require lowering of the amount of pollution and reduction in dependency on fossil fuels. Wastewater management is an area in which these two goals can be addressed simultaneously. As a result, due to limitations of availability of freshwater and ever-increasing energy cost, there has been a paradigm shift recently, from disposing of waste to using it as a resource. While treating wastewater, recovery of energy and valuable materials might reduce the cost of treatment, and to some extent reduce our dependence on fossil fuels. The major limitations to field-scale application of MFCs for treatment of wastewater are that their power density is still relatively low, fabrication cost of this device is higher and the technology is still in the laboratory phase. However, each MFC unit produce limited voltage (0.6–0.8 V); however, by linking several MFCs together, the voltage can be increased or by putting them in parallel connections, current can be integrated, which can be converted to usable form after employing power management system to power the devices.

To make wastewater treatment in MFCs economically viable, the construction and operating costs must be reduced. For enhancing power output by reducing overpotential, the requirement for expensive noble metal catalysts on electrodes, and soluble or electrode-bound electron shuttles are two important cost elements that are being addressed by present researchers. Single-chamber MFCs with air-cathode circumvent the cost of aeration. As compared to other anaerobic treatment processes, MFC also offers advantage of similar organic matter removal efficiency with additional possibility of nutrient and metal removal and recovery. In a nutshell, the possibility of direct conversion of oxidizable organic matter present in wastewater to bio-electricity is exciting, however, fundamental understanding of the microbiology of electrogens and further development of the technology by optimizing geometrical shape, selection of low-cost material for fabricating different components and development of non-noble efficient electrode catalyst, may be in the form of metal composite, are required to make this technology ready for field application.

Acknowledgements The authors acknowledge the grant received from Department of Science and Technology, Government of India (File No. DST/INT/UK/P-101/2014), to undertake this work.

References

Aelterman P, Rabaey K, Pham HT, Boon N, Verstraete W (2006) Continuous electricity generation at high voltages and currents using stacked microbial fuel cells. Environ Sci Technol 40(10):3388–3394

Ahmed J, Yuan Y, Zhou L, Kim S (2012) Carbon supported cobalt oxide nanoparticles–iron phthalocyanine as alternative cathode catalyst for oxygen reduction in microbial fuel cells. J Power Sources 208:170–175

Ahn Y, Logan BE (2010) Effectiveness of domestic wastewater treatment using microbial fuel cells at ambient and mesophilic temperatures. Biores Technol 101(2):469–475

Angenent LT, Karim K, Al-Dahhan MH, Wrenn BA, Domíguez-Espinosa R (2004) Production of bioenergy and biochemicals from industrial and agricultural wastewater. Trends Biotechnol 22 (9):477–485

Aulenta F, Canosa A, Majone M, Panero S, Reale P, Rossetti S (2008) Trichloroethene dechlorination and H_2 evolution are alternative biological pathways of electric charge utilization by a dechlorinating culture in a bioelectrochemical system. Environ Sci Technol 42 (16):6185–6190

Behera M, Jana PS, More TT, Ghangrekar M (2010) Rice mill wastewater treatment in microbial fuel cells fabricated using proton exchange membrane and earthen pot at different pH. Bioelectrochemistry 79(2):228–233

Butler CS, Clauwaert P, Green SJ, Verstraete W, Nerenberg R (2010) Bioelectrochemical perchlorate reduction in a microbial fuel cell. Environ Sci Technol 44(12):4685–4691

Chatterjee P, Ghangrekar MM, Rao S (2016) Low efficiency of sewage treatment plants due to unskilled operations in India. Environ Chem Lett 1–10

Chen S, Chen B (2013) Net energy production and emissions mitigation of domestic wastewater treatment system: a comparison of different biogas-sludge use alternatives. Biores Technol 144:296–303

Cheng S, Liu H, Logan BE (2006) Increased performance of single-chamber microbial fuel cells using an improved cathode structure. Electrochem Commun 8(3):489–494

Cheng S, Xing D, Call DF, Logan BE (2009) Direct biological conversion of electrical current into methane by electromethanogenesis. Environ Sci Technol 43(10):3953–3958

Chung J, Tian DJ, Lee JS, Lim HS, Jun HB (2015) Separation of organic matter from domestic sewage in a UASB-ABF system with anoxic bio-flocculation. Desalin Water Treat

CPCB (2013) Performance evaluation of sewage treatment plants under NRCD. Central Pollution Control Board

Cusick RD, Logan BE (2012) Phosphate recovery as struvite within a single chamber microbial electrolysis cell. Biores Technol 107:110–115

De Schamphelaire L, Boeckx P, Verstraete W (2010) Evaluation of biocathodes in freshwater and brackish sediment microbial fuel cells. Appl Microbiol Biotechnol 87(5):1675–1687

Du Z, Li H, Gu T (2007) A state of the art review on microbial fuel cells: a promising technology for wastewater treatment and bioenergy. Biotechnol Adv 25(5):464–482

Dutta PK, Keller J, Yuan Z, Rozendal RA, Rabaey K (2009) Role of sulfur during acetate oxidation in biological anodes. Environ Sci Technol 43(10):3839–3845

Ghangrekar MM, Shinde VB (2007) Performance of membrane-less microbial fuel cell treating wastewater and effect of electrode distance and area on electricity production. Biores Technol 98(15):2879–2885

Ghangrekar MM, Shinde VB (2008) Simultaneous sewage treatment and electricity generation in membrane-less microbial fuel cell. Water Sci Technol 58:37–43
Gregory KB, Lovley DR (2005) Remediation and recovery of uranium from contaminated subsurface environments with electrodes. Environ Sci Technol 39(22):8943–8947
Helder M, Strik D, Hamelers H, Kuhn A, Blok C, Buisman C (2010) Concurrent bio-electricity and biomass production in three plant-microbial fuel cells using *Spartina anglica*, *Arundinella anomala* and *Arundo donax*. Biores Technol 101(10):3541–3547
Hindatu Y, Annuar MSM, Gumel AM (2017) Mini-review: anode modification for improved performance of microbial fuel cell. Renew Sustain Energy Rev 73:236–248
Jadhav DA, Ghangrekar MM (2016) Effective ammonium removal by anaerobic oxidation in microbial fuel cells. Environ Technol (United Kingdom) 36(6):767–775
Jadhav D, Ghadge A, Mondal D, Ghangrekar M (2014) Comparison of oxygen and hypochlorite as cathodic electron acceptor in microbial fuel cells. Biores Technol 154:330–335
Kelly PT, He Z (2014) Nutrients removal and recovery in bioelectrochemical systems: a review. Biores Technol 153:351–360
Kim BH, Gadd GM (2008) Bacterial physiology and metabolism. Cambridge University Press
Kim Y, Logan BE (2011) Series assembly of microbial desalination cells containing stacked electrodialysis cells for partial or complete seawater desalination. Environ Sci Technol 45 (13):5840–5845
Kim BH, Chang IS, Gil GC, Park HS, Kim HJ (2003) Novel BOD (biological oxygen demand) sensor using mediator-less microbial fuel cell. Biotech Lett 25(7):541–545
Kim JR, Dec J, Bruns MA, Logan BE (2008) Removal of odors from swine wastewater by using microbial fuel cells. Appl Environ Microbiol 74(8):2540–2543
Lee DJ, Liu X, Weng HL (2014) Sulfate and organic carbon removal by microbial fuel cell with sulfate-reducing bacteria and sulfide-oxidising bacteria anodic biofilm. Biores Technol 156:14–19
Lens PNL, Visser A, Janssen AJH, Hulshoff Pol LW, Lettinga G (1998) Biotechnological treatment of sulfate-rich wastewaters. Criti Rev Environ Sci Technol 28(1):41–88
Logan BE (2008) Microbial fuel cell. Wiley, Hoboken
Lu N, Zhou S, Zhuang L, Zhang J, Ni J (2009) Electricity generation from starch processing wastewater using microbial fuel cell technology. Biochem Eng J 43(3):246–251
Luo H, Jenkins PE, Ren Z (2010) Concurrent desalination and hydrogen generation using microbial electrolysis and desalination cells. Environ Sci Technol 45(1):340–344
Martin E, Tartakovsky B, Savadogo O (2011) Cathode materials evaluation in microbial fuel cells: a comparison of carbon, Mn_2O_3, Fe_2O_3 and platinum materials. Electrochim Acta 58:58–66
Min B, Kim J, Oh S, Regan JM, Logan BE (2005) Electricity generation from swine wastewater using microbial fuel cells. Water Res 39(20):4961–4968
Morris JM, Jin S (2007) Feasibility of using microbial fuel cell technology for bioremediation of hydrocarbons in groundwater. J Environ Sci Health Part A 43(1):18–23
Nelson DL, Lehninger AL, Cox MM (2008) Lehninger principles of biochemistry. Macmillan
Nevin KP, Richter H, Covalla S, Johnson J, Woodard T, Orloff A, Jia H, Zhang M, Lovley D (2008) Power output and columbic efficiencies from biofilms of *Geobacter sulfurreducens* comparable to mixed community microbial fuel cells. Environ Microbiol 10(10):2505–2514
Oh S, Logan BE (2005) Hydrogen and electricity production from a food processing wastewater using fermentation and microbial fuel cell technologies. Water Res 39(19):4673–4682
Pant D, Singh A, Van Bogaert G, Olsen SI, Nigam PS, Diels L, Vanbroekhoven K (2012) Bioelectrochemical systems (BES) for sustainable energy production and product recovery from organic wastes and industrial wastewaters. RSC Adv 2(4):1248–1263
Pradhan H, Ghangrekar MM (2014a) Multi-chamber microbial desalination cell for improved organic matter and dissolved solids removal from wastewater. Water Sci Technol 70(12):1948–1954
Pradhan H, Ghangrekar MM (2014b) Organic matter and dissolved salts removal in a microbial desalination cell with different orientation of ion exchange membranes. Desalin Water Treat

Rabaey K (2010) Bioelectrochemical systems: from extracellular electron transfer to biotechnological application. IWA publishing
Rabaey K, Verstraete W (2005) Microbial fuel cells: novel biotechnology for energy generation. Trends Biotechnol 23(6):291–298
Rabaey K, Rozendal RA (2010) Microbial electrosynthesis—revisiting the electrical route for microbial production. Nat Rev Microbiol 8(10):706–716
Rabaey K, Van De Sompel K, Maignien L, Boon N, Aelterman P, Clauwaert P, De Schamphelaire L, Pham HT, Vermeulen J, Verhaege M, Lens P, Verstraete W (2006) Microbial fuel cells for sulfide removal. Environ Sci Technol 40(17):5218–5224
Ray SG, Ghangrekar MM (2016) Biodegradation kinetics of thin-stillage treatment by *Aspergillus awamori* and characterization of recovered chitosan. Appl Microbiol Biotechnol 100(4):1955–1965
Rozendal RA, Leone E, Keller J, Rabaey K (2009) Efficient hydrogen peroxide generation from organic matter in a bioelectrochemical system. Electrochem Commun 11(9):1752–1755
Sajana TK, Ghangrekar MM, Mitra A (2014) Effect of presence of cellulose in the freshwater sediment on the performance of sediment microbial fuel cell. Biores Technol 155:84–90
Sajana TK, Ghangrekar MM, Mitra A (2017) Influence of electrode material on performance of sediment microbial fuel cell remediating aquaculture water. Environ Eng Manage J 16(2):421–429
Sato N, Okubo T, Onodera T, Ohashi A, Harada H (2006) Prospects for a self-sustainable sewage treatment system: A case study on full-scale UASB system in India's Yamuna River Basin. J Environ Manage 80(3):198–207
Schamphelaire LD, Bossche LVD, Dang HS, Höfte M, Boon N, Rabaey K, Verstraete W (2008) Microbial fuel cells generating electricity from rhizodeposits of rice plants. Environ Sci Technol 42(8):3053–3058
Shi X, Feng Y, Wang X, Lee H, Liu J, Qu Y, He W, Kumar S, Ren N (2012) Application of nitrogen-doped carbon powders as low-cost and durable cathodic catalyst to air–cathode microbial fuel cells. Biores Technol 108:89–93
Steinbusch KJ, Hamelers HV, Schaap JD, Kampman C, Buisman CJ (2009) Bioelectrochemical ethanol production through mediated acetate reduction by mixed cultures. Environ Sci Technol 44(1):513–517
Sun J, Hu Y, Bi Z, Cao Y (2009a) Improved performance of air-cathode single-chamber microbial fuel cell for wastewater treatment using microfiltration membranes and multiple sludge inoculation. J Power Sources 187(2):471–479
Sun M, Sheng G-P, Mu Z-X, Liu X-W, Chen Y-Z, Wang H-L, Yu H-Q (2009b) Manipulating the hydrogen production from acetate in a microbial electrolysis cell–microbial fuel cell-coupled system. J Power Sources 191(2):338–343
Van Haandel AC, Lettinga G (1994) Anaerobic sewage treatment: a practical guide for regions with a hot climate. John Wiley & Sons, Singapore
Wang H, Ren ZJ (2013) A comprehensive review of microbial electrochemical systems as a platform technology. Biotechnol Adv 31(8):1796–1807
Wang X, Feng Y, Lee H (2008) Electricity production from beer brewery wastewater using single chamber microbial fuel cell. Water Sci Technol 57(7):1117–1121
Zhang X, Cheng S, Wang X, Huang X, Logan BE (2009) Separator characteristics for increasing performance of microbial fuel cells. Environ Sci Technol 43(21):8456–8461
Zhao F, Rahunen N, Varcoe JR, Chandra A, Avignone-Rossa C, Thumser AE, Slade RCT (2008) Activated carbon cloth as anode for sulfate removal in a microbial fuel cell. Environ Sci Technol 42(13):4971–4976

Chapter 10
Remediation of Industrial Effluents

Ashutosh Rai, Pankaj Singh Chauhan and Shantanu Bhattacharya

Abstract Environmental pollution has become the greatest problem of mankind in the recent times. The level of pollutants especially in water is increasing day by day causing a grave danger to all the living entities. The industrial effluent has been identified as a major polluting agent of the water resources. Various industries utilise water in cooling, processing, transportation and heat treatment. The water gets mixed with various harmful chemicals during these processes and forms a harmful effluent which is directly discharged to rivers by these industries. This effluent contaminates the water resources and soils affecting the human, animal and plant life in a large scale. Hence, the remediation of industrial effluent before discharging into the environment becomes essential. Various physical, chemical and biological processes are being utilised to provide a complete remediation or remediation to an extent where it could be discharged safely or could be reused. This chapter involves a complete explanation of all the processes with their advantages as well as disadvantages. Many of the above-mentioned industrial applications discharge complex organic compounds in the form of effluent which is very difficult to degrade by using conventional methods such as photoreaction. Hence, a catalyst-based photoreaction called photocatalysis is utilised by many researchers for the degradation of such complex organic compounds, present in wastewater. The metal oxide semiconductors (MOSs) such as titanium dioxide (TiO_2), tin dioxide (SnO_2) and zinc oxide (ZnO) have been used as catalyst for the dye degradation of industrial wastewater.

Keywords Photocatalysis · Wastewater remediation · Industrial effluents
Metal oxide semiconductors

A. Rai · P. S. Chauhan (✉) · S. Bhattacharya
Microsystems Fabrication Lab, Department of Mechanical Engineering,
Indian Institute of Technology, Kanpur, India
e-mail: pankajs@iitk.ac.in

S. Bhattacharya
Design Program, Indian Institute of Technology, Kanpur, India

S. Bhattacharya et al. (eds.), *Water Remediation*, Energy, Environment, and Sustainability, https://doi.org/10.1007/978-981-10-7551-3_10

10.1 Introduction

With the industrial revolution, environmental pollution has become menace to the world. Environmental pollution is one of the greatest concerns in today's world of which industries play a keen role in eroding our natural resources. Industries acquire huge amount of water while processing and then discharge these effluents directly into the habitat. These effluents are usually discharged to the ocean leading to the destruction of the aquatic life due to the chemicals present in the effluent, which are usually toxic and can cause serious health issues. These chemicals also do exhibit physical properties making the water aesthetically unpleasant. The major physical characteristics of these discharged effluents are its heavy metal content, colour as well as odour. The solids present in the wastewater can have dual nature, i.e. insoluble and suspended form. One simple method to measure these suspended solids is by drying and weighing the leftovers by filtering. Volatile solids (mostly having organic content) are easily removed by burning them off using ignition. Organic matters include protein, carbohydrate and fats. On heating at elevated temperatures, some of the organic matters burn incompletely and some of the inorganics break down.

It has been documented in 2004 by World Health Organization that approximately 17% of the global population did not get safe drinking water of which most of the population reside in countries including China, India, Middle East and Africa as well as about 42% of sub-Saharan Africa lacked drinking water. At the end of 2008, around 884 million people in the world were deprived of better sources of drinking water. It has been forecasted that more than 47% of the world's population would be facing critical water trouble in the year 2030 (Mathew Akinloye and Olatunde Olubanjo 2014). So there is a huge need to save water.

Industries are advised to treat wastewater effluent before discharging as per the norms set by the government. There are some general standards for discharge of effluents and wastewater generation. TDS (total dissolved solids), pH, BOD (biochemical oxygen demand), COD (chemical oxygen demand), TOC (total organic carbon) are some of the parameters to be controlled. There are some methods which are commonly utilised which include coagulation, flocculation, filtration, sedimentation. All these methods do have their pros and cons. The most significant is the textile industries which discharge both organic dyes and inorganic particulates in a heavy proportion to clean marine. The degradation rate of these effluents in visible light can be catalysed by the implementation of semiconductor as well.

10.2 Pollutants of Different Industries

Sources of industrial wastewater pollutants depending upon the type of industry are described as follows:

10.2.1 Textile Industries

Textile industries consume a lot of water while fabric processing; for processing 1 kg of fibre, approximately 200 L of water is required. A large number of chemical products are used in these textile industries for the cleaning as well as dyeing of work materials. Released effluents from these textile industries contain huge amounts of hazardous contaminants and heavy metals. Sometimes, the formation of some toxic aromatic amines occurs in aromatic media (Jin et al. 2007). Wastewater released from textile industries contains aqueous waste and organic dye. This aqueous waste could be organic or inorganic pollutants. When this wastewater is discharged into the surface water like rivers, ponds, and canals, it adds some colour to the water because of the presence of organic dye. This addition of colour in the water makes the water look aesthetically obnoxious and even blocks or reduces the entrance of sunlight through the water and affects photosynthesis which leads to dying of aquatic species due to lack of oxygen. The presence of other organic and inorganic pollutants leads to high BOD, making the water more toxic and harmful to use (Pereira 2012). Currently, more than 10,000 different dyes are available commercially and more than 7×10^5 tons per year are produced around the world to meet the population demand. Because of inadequate dyeing procedure in textile industry, 10–25% of the dye produced is lost during the dyeing process, and 2–20% of the dyes are usually not utilised and are released as effluents directly (Carmen and Daniel 2012). It becomes a noticeable fact that even a very small amount of dye in water (less than 1 ppm for some dye) is highly undesirable (Zonoozi et al. 2008). Synthetic dye is also used in the industry, widely. Above-mentioned data suggest that there is a release of about 2,80,000 tons of textile dyes every year all over the world (Jin et al. 2007). Dye-containing effluent when released into the environment is objectionable because of colour, and many of the dyes discharged are toxic and their breakdown product can be carcinogenic and mutagenic to the living entities. It causes severe health issues like skin irritation or allergies due to contact dermatitis, hypersensitive eye reaction, respiratory disorders and also irritation to skin. Groundwater quality is also being affected, and soil productivity is degrading due to untreated effluent discharge to different environmental sources (Carmen and Daniel 2012). Colour removal from the textile effluent is the strenuous task. Breakdown of dyes, mainly azo dyes, comprising almost 70% of all dyes used, is difficult because of composite structure and synthetic nature (Jin et al. 2007). Table 9.1 describes different polluting characteristics present at different processing steps in cotton textile industry.

Table 10.1 Wastewater characteristics at different stages of processing cotton textiles and permissible limits to discharge into water bodies (Reprinted with permission of Chapagain et al. 2006)

Process	Wastewater volume (m^3/ton)	Pollutants (kg/ton)			
		BOD	COD	TSS	TDS
Wet processing	360	32	123	25	243
Bleaching	30	5	13		28
Dyeing	142	6	24		180
Printing	188	21	86	25	35
Finishing	136	6	25	12	17
Total	496	38	148	37	260
Permissible Limits (mg/L)		50	250	50	

10.2.2 Power Plants and Nuclear Industries

Due to the huge demand of industry, agricultural sectors as well as human population and other sectors in developing country like India, the growth of power and the development of the country are directly related. About 70–80% of the total power demand is coped with coal-based thermal power stations. Rough estimate points out that about 91,000 m^3 of water is required per day for the generation of 840 MW of power (Kamdi et al. 2012). These power stations utilise huge amount of water in each and every step starting from steam generation to auxiliary cooling, thereby discharging huge amount of contaminants in the form of suspended solids and other heavy metals including chlorine, zinc, chromium and phosphate. Waste released will affect both the soil and groundwater, thereby leading to health issues. Some of the factors that are affected are pH, COD, suspended solids, TDS and heavy metals.

10.2.3 Iron and Steel Industry

India is one of the major producers of crude steel. In 2016, China remained the largest producer of 808 MT (million tonnes), followed by Japan (105 MT) and India (96 MT) (An Overview of Steel Sector). It has been concluded that about 250–500 m^3/hr is released as wastewater for the manufacture of about 1 million tonnes of ingot steels (Geny and Dohen 1972). It can be seen that it is a huge amount; hence, there is a huge demand for the water to be reused. The effluents discharged can be seen from Table 9.2 (Geny and Dohen 1972).

Table 10.2 Different types of waste materials added at different positions in plants (Geny and Dohen 1972)

Plants	Waste material added in plant
Coke oven plant	Phenols, cyanides, various tars, ammonia
Burden preparation	Suspended solids and dissolved salts
Cast iron manufacture	Suspended solids, cyanides, phenols, ammonia, granulated slag, lime
Steel manufacture	Suspended solids, lime
Hot rolling	Lime, suspended solids, oils
Cold rolling	Oils, suspended solids

10.2.4 Pulp and Paper Industry

Due to the rapid increase in population worldwide, there is a huge demand of paper and pulp for both industrial developments and human requirements. The water usage for its production is quite high ranging between 20,000 and 60,000 gallons per ton of product (Pokhrel and Viraraghavan 2004). Table 9.3 shows the major contaminants discharged from pulp and paper industry.

These dyes which are discharged from different industries, i.e. pharmaceutical industries, paper industry, tannery and bleaching industry, constitute various types of organic as well as inorganic pollutants. Chemical industries utilise a huge amount of water but that is less compared to that of textile industries.

The recycling has thus been recommended due to the huge demand and less availability. Nowadays, there is very keen interest in the remediation of the effluent due to high levels of contamination in dyeing, i.e. dyes and their products, pigments, organic chemicals as well as heavy metals. These include detergents, VOCs, sizing agents and fixing agents, inorganics, latex, dyes and a wide variety of special chemicals. Depending upon various industries, the contaminants in the effluents are

Table 10.3 Contaminants from pulp and paper industry (Reprinted with permission of Pokhrel and Viraraghavan 2004)

Manufacturing system	Contaminants
Wood preparation	Dirt, grit, lignin, suspended solids, BOD, etc.
Digester house	Resins, fatty acids, colour, volatile organic compounds (methanol, acetone, isopropyl alcohol, etc.), BOD, COD
Pulp washing	High pH, BOD, COD, suspended solids, dyes
Pulp bleaching	Dissolved lignin, carbohydrate, dyes, AOX, inorganic chlorine compounds such as chlorate, organochlorine compounds such as dioxins, VOCs (volatile organic compounds) such as acetone, methyl chloride, carbon disulphide, chloroform, chloromethane, trichloromethane
Papermaking	Suspended particulate waste, organic compounds, azo dyes, acetone

different. Despite all these factors, there exhibit very few industries which treat these effluents before discharging it in the sea.

The motivations for the removal of these contaminants from the effluent are as follows:

1. There is a fixed capacity of surface water.
2. To minimise the decrease in the aquatic species.
3. To safeguard humans from waterborne diseases.

The selection and design of treatment facilities is based on a study of the following:

1. The TDS pH, BOD, COD, TOC of wastewater.
2. The standards that must be fulfilled to discharge these effluents in environment or for reuse of this effluent.

There are two broad subdivisions of the chemical characteristics of these effluents, i.e. inorganic and organic. Volatile organic compounds are the major organic contaminants which consist of terpenes, phenols, acetone, methanol, chloroform, alcohols, etc. To take care of these toxic substances, the techniques are subdivided into three major methods, i.e. chemical treatment, physical treatment and biological treatment. The chemical method includes oxidation from various oxidising agents. Physical treatment refers to Adsorption from various materials like Granulated Activated Carbon, Wood chips, Membrane filtration from Reverse osmosis, Ion Exchange method etc. While biological treatment is carried out with the help of micro-organisms including fungi, bacteria etc. using aerobic or anaerobic respiration.

Daily enormous amount of these effluents are being discharged into the surface, thereby harming the aquatic species as well as human health. So there is a need to protect the surface water to save the aquatic life, maintain the aesthetic value of the water and reduce the harmful effect of pollutants on human health. To remove the pollutants from the wastewater effluent, there is a need to treat this effluent before discharging.

10.3 Remediation Methods for Pollutants

Due to the growth in the industries, the amount of heavy metals and organic compounds produced are rising and approximately 10,000 new organic compounds are being added every year. In the future, with the goal of pollution prevention, all efforts have to be made for industrial dischargers to estimate the environmental impacts of the new compounds that can enter the wastewater effluent, before it is being accepted for use. If a compound cannot be effectively treated with existing technology, it should be banned. Today, 70% of available water in India is contaminated and about 66% of the diseases in India are waterborne diseases (Robinson et al. 2001). Water quality parameters like acidity, nitrogen, alkalinity,

BOD (biological oxygen demand), COD (chemical oxygen demand), dissolved oxygen, heavy metals, conductivity, pH, TDS (total dissolved solids), turbidity are required to get evaluated before and after treatment. There are several methods which are in practice to treat industrial wastewater. Different water treatment plants are implemented by the industry to remove a precise amount of pollutants from the effluent and meet general standards of effluent discharge. These treatment units imply physical, chemical and biological techniques to remove pollutants from the wastewater. If these water treatment plants are operated in a genuine way, then it can manage most of the pollutants. Some of the toxic pollutants cannot be treated like cyanides and excessive amount of heavy metals by these processes (Munter 2003). Table 9.3 shows the methods used for controlling different water quality parameters.

10.3.1 Removal of Dye Colour from Effluent

Numerous methods are being practiced to treat industrial wastewater before discharging it into ecosystem. Presently, miscellaneous chemical, physical and biological methods are used to remove colour from textile effluent, but these methods have some pros and cons.

Though physical methods are successful in removing colour from the water, it is not efficient to degrade dye molecules and requires safe disposal of concentrated sludge (Robinson et al. 2001). Chemical methods are also able to remove dye from the effluent, but create another disposal problem with large accumulation of concentrated sludge. And with heavy amount of chemical requirement, there may be a possibility of secondary disposal problem. Biological methods are effective enough to completely mineralise organic impurities and to remove BOD, COD and suspended solids and considered as environmental-friendly (Robinson et al. 2001).

Despite, successful dye removing characteristics, physical and chemical methods are not commercially striking. However, biological methods require less cost and are used widely to treat dye effluent.

10.3.2 Physical Treatment

Some of the physical treatment methods are as follows:

10.3.2.1 Coagulation and Flocculation

Coagulation is a path followed to eliminate the stability of any suspension in a solution. The process in which these formed particles are induced to come together to form a particular larger sludge is termed as flocculation. Small particles in

Fig. 10.1 Coagulated sludge from industrial wastewater

effluent contain a negative charge, which inhibits particles to stick to each other. Therefore, when a coagulant is added, this reduces the negative charge of particles by providing a positive charge. Once the reduction of charge occurs, particles become free to form a floc (Bratby 1980). Flocculation is a process of stirring water in the presence of coagulant with accurate velocity to form flocs and settle down due to gravity or neutralise the floc. Further filtration/precipitation is required to remove this settled floc/sludge. Generally used coagulants are alum, ferric chloride, etc. The process is economically feasible because of its low input but the generation of sludge results in higher disposal cost as well as environmental constraints. Figure 10.1 shows the example of coagulated sludge from industrial wastewater.

10.3.2.2 Adsorption

Activated carbon filter or charcoal is used to treat effluents; it is advantageous because of its adsorbing nature of both the organic and inorganic contaminants onto its surface. Adsorption has also attracted researchers towards it because of its efficiency in the removal of pollutants which are too stable for regular methods. Processed carbon having small and low volume pores is termed as activated carbon. Due to its presence of pores, there is an increase in the surface area leading to adsorption of impurities on to its surface due to its ultra-porous formation. It has extremely high surface area of about 500–1500 m^2/gm (Kamdi et al. 2012). Some adsorbates like silica gel, corn cobs, rice hulls, fly ash and coal mixture make a high-quality adsorbent which can remove contaminants even from diluted solution and produce a high grade of treated effluent without resulting into any harmful product. It has few major disadvantages like adsorbent quality starts degrading after a period of time due to its repeated use and comparatively high capital cost (Robinson et al. 2001).

10.3.2.3 Granular Activated Carbon

Granular activated carbon is usually made from organic materials such as: wood, lignite and coal etc., and contains a high amount of carbon. Activated carbon commonly adsorbs both natural and synthetic organic compounds as well as compounds providing taste and odour (Magic-Knezev and van der Kooij 2004). Activated carbon is highly porous in nature and its granulated form has large surface area, which facilitates higher adsorption of contaminants . Granular activated carbon is utilised usually for the removal of pharmaceutical impurities.

10.3.2.4 Peat

Peat is also one of the best adsorbents because of its cellular structure. It could adsorb the transition metals as well as the polar organic compounds from the dyed effluents (Robinson et al. 2001). It also does not require any activation. Used peat could be utilised further by burning it.

10.3.2.5 Wood Chips

It is having best adsorption capacity for acidic dyes. It is disadvantageous because of its longer contact time.

10.3.2.6 Universally Available Materials

There are various materials which are widely available as well as could be utilised for the process which includes natural clay, rice hulls (Robinson et al. 2001). They are very economical for the removal of dye.

10.3.2.7 Reverse Osmosis

Reverse osmosis is defined as the movement of solute from higher concentration to lower concentration through a semi-permeable membrane by applying a pressure exceeding the osmotic pressure. It is a purification technique which utilises semi-permeable membrane which is having layered and weblike structure in which water follows a tortuous path when a pressure is applied. It is utilised for desalination and the removal of ions and molecules (Greenlee et al. 2009). There is a need to improve the membrane technology, so that heavy metals (As, Al, Hg etc.) can be filtered. Also, its domestic use in urban as well as rural areas should be improved by reducing the cost. It has many advantages which include temperature resistant and microbial attack resistant (Robinson et al. 2001).

10.3.2.8 Ion Exchange Method

In this process, the effluents are passed through ion exchange resin until there is saturation at available exchange sites as shown in Fig. 10.2. It could remove both the cationic and anionic dyes. Ion exchange method leads to the removal of all ions from solution or separates the substances. It has resulted in the removal of heavy metal ions such as lead, mercury, cadmium, nickel, vanadium, chromium, copper and zinc. Usually, these ions are precipitated in the form of metal oxides, xanthogenates or sulphides (Dąbrowski et al. 2004). It is advantageous because of no loss of absorbent on regeneration, and the solvent can also remove soluble dyes (Robinson et al. 2001). The main disadvantages are the disposal issues of the huge amount of sediments which contain heavy metal ions formed and organic solvents are not economical.

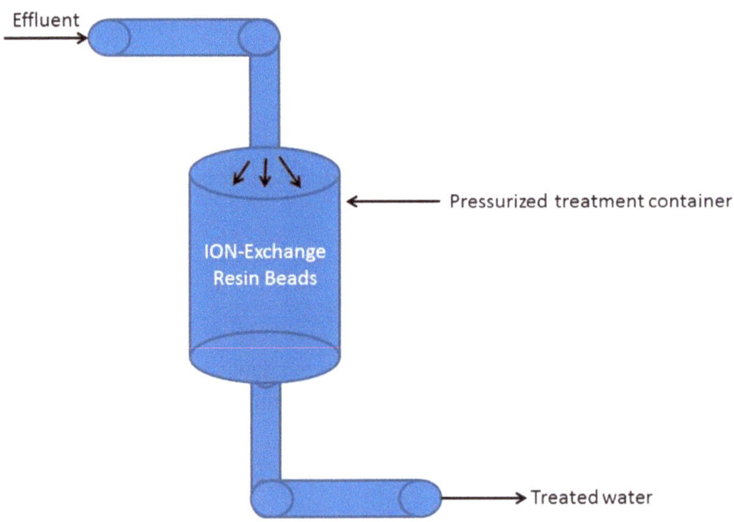

Fig. 10.2 Ion exchange process

10.3.3 Biological Treatment

10.3.3.1 Aerobic and Anaerobic Treatment

In aerobic treatment, bacteria or micro-organisms help in the degradation of the organic impurities in the presence of oxygen/air and transform these impurities into carbon dioxide, water and biomass as shown in Fig. 10.3. All these processes are being performed in a bioreactor by maintaining certain operating conditions. Similarly, in anaerobic treatment, bacteria or micro-organisms which do not need oxygen/air to degrade organic impurities are introduced.

10.3.3.2 Degradation by Micro-organism

Micro-organisms including fungus have been utilised in the recent decades for the removal of dyes from various dyed effluents. White-rot fungi are the organisms having the capability to degrade lignin (Barr and Aust 1994). White-rot fungi degrade the dyes by utilising enzymes, such as lignin peroxides. Azo dyes (readily used dyes) are not readily degraded by micro-organisms but can be degraded by Phanerochaete chrysosporium. Phanerochaete chrysosporium (white-rot fungi) are also utilised for xenobiotic degradation, degrading dioxins and various other chloro-organics (Barr and Aust 1994). *Aspergillus fumigatus* XC6 was usually collected with the help of rice straw. It works at a pH range of 3–8, and it has been

Fig. 10.3 Aeration tank showing flow of dissolved oxygen

proved that *A. fumigatus* XC6 was maximum efficient when the dye is a source of carbon and nitrogen, with pH of 3 after 48 h of incubation (Jin et al. 2007). Other than fungi, it has been reported that bacteria also have the ability to metabolise azo dyes. Azo dyes are not metabolised readily under aerobic conditions but it has been reported that there is an ability of Pseudomonas strain to aerobically degrade certain azo dyes. There have been some additional processes for the degradation of the azo dyes. Many researchers have found bacteria to be efficient in reducing a variety of sulfonated as well as non-sulfonated azo dyes anaerobically. High molecular-sized and highly charged sulfonated and polymeric azo dyes are not likely to pass cell membrane (Keck et al. 1997).

Although it has been shown that white-rot fungi could be utilised for decolourising dyes in liquid fermentation, enzyme production is unreliable due to the unfavourable environment of liquid fermentation. They are more likely to work in solid-state fermentation (Robinson et al. 2001).

10.3.4 Chemical Method

10.3.4.1 Nitrification and Denitrification

Nitrate is regarded as one of the most undesirable components in drinking water which is present in groundwater mainly because of the fertilisers being utilised as well as the waste from humans. Nitrification is a process of conversion of ammonia (highly toxic) to nitrate. Ammonia is first converted into nitrite and finally to nitrate. Denitrification is a process in which the formed nitrate is converted into nitrogen. This method does have a limitation because of its selective nature.

10.3.4.2 Chlorination

Chlorine is highly toxic in nature and hence could be utilised to kill various types of microbes. When dissolved in water, it mainly results in two components which include hypochlorous acid and hydrochloric acid. There are various harmful micro-organisms, which can cause diseases such as: cholera, typhoid, dysentery etc., can be rectified by chlorination. Chlorination is not effective for the remediation of an effluent having inorganic impurities. Also, its disinfection by-products are harmful. if present in the treated water.

10.3.4.3 Ozone

The use of ozone had been initiated since 1970s. Its oxidation potential is very high, i.e. 2.07 which is very large as compared to both the chlorine (1.36) and hydrogen peroxide (1.78) (Robinson et al. 2001). Ozone being the strongest oxidising agent (removes metals like iron, manganese as well as inorganic substances like cyanides, sulphides) is used widely in the remediation process. There is no formation of residue or sludge, and the quantity of ozone required depends on its colour and residual COD (Ince and Gönenç 1997). Ozone completely mineralises the natural organic matter (NOM) which is the major reason of the improper taste and odour of the drinkable water. It is widely used for decolouration, conversion of COD to BOD as well as inactivation of viruses. It has widely replaced chlorine purification system because of no harmful by-products as was visualised in case of chlorine. Ozone has a major advantage of its application in gaseous state, thereby not increasing the volume of the treated effluent. Dyes in the chromosphere groups having conjugated double bonds could be broken down to smaller molecules, thus having a reduction in colour (Peralta-Zamora et al. 1999). After its breakdown, ozonation process could be utilised because these smaller molecules tend to have an increase in the toxic properties. Ozonation also has a short half-life (20 min) which leads to an increase in its cost. This time is also shortened if there is a presence of dyes in the effluent (Robinson et al. 2001).

10.3.4.4 Cucurbituril

Cucurbituril is a cyclic polymer of glycoluril and formaldehyde. It is named so because of its pumpkin structure. It has been shown by researchers that it has an extraordinary good sorption capacity for various types of textile dyes (Robinson et al. 2001).

10.3.4.5 Electrochemical Destruction

This is one of the recent techniques utilised. It is most advantageous because of no chemical usage as well as no sludge formation taking place. The end products are not harmful (Robinson et al. 2001).

10.3.4.6 Photochemical Method

In all the above remediating methods, there are some drawbacks which must be sorted out. The removal of dye by its degradation is the best viable method present. There are various advantages of this method which include the following:

1. This process is not compound-specific, hence, any hazardous organic chemicals, fuel hydrocarbon and dye could be remediated.

2. The process is resistant to toxicity as neither the initials are toxic nor the by-products are toxic.
3. The process could be applied to both liquid and gaseous currents.
4. The requirement for the process to perform well is solar energy which is widely available.
5. The process is economically viable as compared to any other process.

There are various metal oxide semiconductors available which could be utilised for the degradation of the dye by photocatalysis which include titanium dioxide (TiO_2), zinc oxide (ZnO) and vanadium pentoxide (V_2O_5). (Chauhan and Bhattacharya 2016). MOSs catalyse the photoreaction by creating electron–hole pair after the absorption of small amount of energy which depends upon their band gap. These electron–hole pairs form hydroxyl radicals which act as a strong oxidising agent for complex organic dyes and degrade them. Among these MOSs, zinc oxide (ZnO) has been found to be advantageous over others due to low cost, capability of absorbing large amount of UV light and variety in shape and size of nanostructure, (Chauhan et al. 2017). The surface defects present in the form of oxygen vacancies in crystal structure of ZnO nanostructures are found to be helpful in efficient dye degradation process. The presence of these defects can be increased by utilising solvothermal method of fabrication of ZnO as reported by the studies. One of such process is shown in Fig. 10.4.

This chapter describes the various aspects and advances in industrial wastewater remediation by using MOSs. The main reason for the dye degradation is the presence of electron (in the conduction band). This electron results in the generation of oxide radical which is highly reactive. The schematic diagram showing the

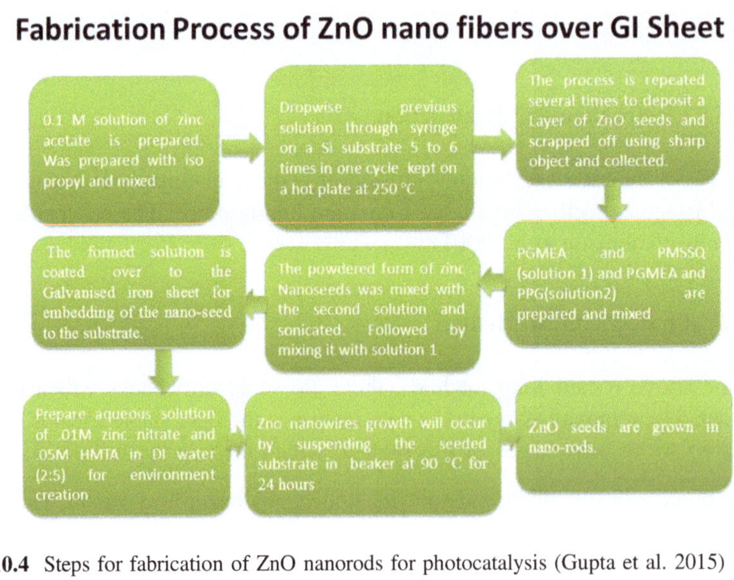

Fig. 10.4 Steps for fabrication of ZnO nanorods for photocatalysis (Gupta et al. 2015)

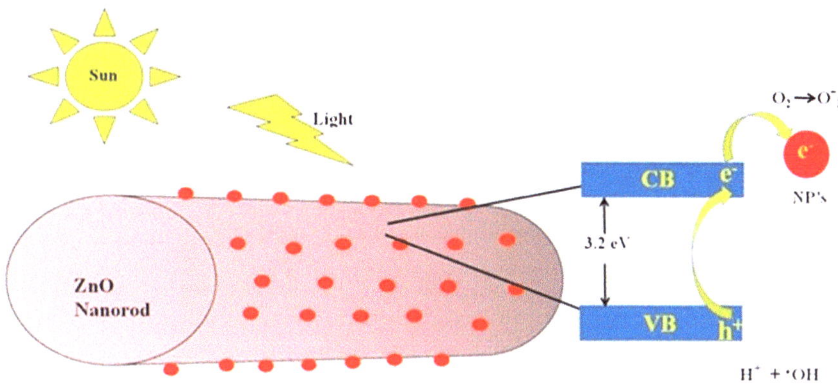

Fig. 10.5 Photocatalytic reaction under sunlight (Chauhan et al. 2017)

photocatalysis reaction taking place at the surface of ZnO nanostructures is shown in Fig. 10.5. The reactions taking place during photocatalysis are as follows:

$$ZnO + h\nu \rightarrow e^-_{CB} + h^+_{\nu B}$$
$$h^+_{\nu B} + H_2O \rightarrow H^+ + OH$$
$$e^-_{CB} + O_2 \rightarrow O_2^-$$
$$O_2^- + \dot{O}H + \text{Rh dye} \rightarrow \text{degraded by-products.}$$

These degraded by-products are then filtered out from the effluents.

10.4 Suggestion for Dye Removal

After the photochemical degradation process, the degraded organic and inorganic by-products are passed through an adsorbate (preferably the granulated activated carbon) so that the leftovers could be captured. To increase the efficiency of the plant, there must be cleaning process to take place at regular intervals. The effluents are then coagulated so that both heavy metal anions and cations can be captured. This coagulation results in the formation of sludge in huge quantity; hence, there must be neat disposal of the formed sludge. It has been suggested that these sludges could be reused in the cement or ceramic industries, because of its properties such as brittleness. Figure 10.6 shows the proposed steps to be performed serial-wise.

10.5 Conclusion

Some of the methods like membrane filtration with reverse osmosis, filtration through cucurbituril are quite efficient but have a limitation of the flow volume; i.e. it can be applied at household applications but when it comes to large water

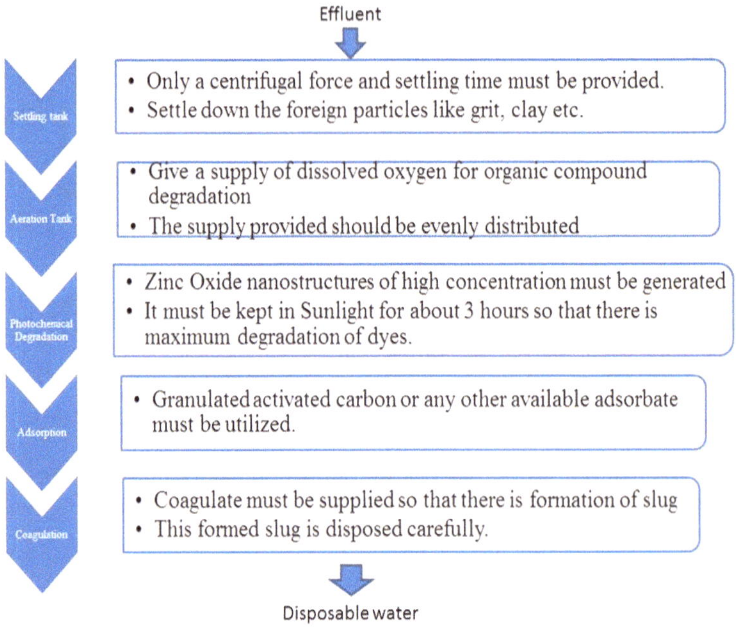

Fig. 10.6 Steps for effluent treatment

treatment plants, they fail to give satisfactory results. Biological activities provide a very bad result when it comes to the degradation of dyes, the main reason being the time period which is very large. Different oxidising agents like chlorine, ozone and hydrogen peroxide could also be utilised but the limiting condition remains the cost aspect. Ozone being the best reducing agent could degrade the dyes but it would be required in large quantity because of the large amount of effluent as well as its short working life. Fungi could be utilised for dealing with the maximum amount of organic impurities and minimum amount of inorganic impurities because of its constraint. Hence, there must be a plan which must involve physio-chemical process for the remediation of the dye. This is followed by the photochemical degradation of the dye which is the most important step as it is not compound-specific, and maximum amount of dye degradation could take place.

References

An Overview of Steel Sector. In: minist. Steel gov. India

Barr DP, Aust SD (1994) Mechanisms white rot fungi use to degrade pollutants. Environ Sci Technol 28:78A–87A. https://doi.org/10.1021/es00051a002

Bratby J (1980) Coagulation and flocculation: with an emphasis on water and wastewater treatment. Filtr Sep 55–58

Carmen Z, Daniel S (2012) Textile organic dyes—characteristics, polluting effects and separation/ elimination procedures from industrial effluents—a critical overview. Org Pollut Ten Years After Stock Conv—Environ Anal Updat. https://doi.org/10.5772/32373

Chapagain AK, Hoekstra AY, Savenije HHG, Gautam R (2006) Valuation of ecosystem goods and services part 1: an integrated dynamic approach. Ecol Econ 60:186–203. https://doi.org/10.1016/j.eco

Chauhan PS, Bhattacharya S (2016) Vanadium pentoxide nanostructures for sensitive detection of hydrogen gas at room temperature. J Energy Environ Sustain 2:69–74

Chauhan PS, Rai A, Gupta A, Bhattacharya S (2017) Enhanced photocatalytic performance of vertically grown ZnO nanorods decorated with metals (Al, Ag, Au, and Au–Pd) for degradation of industrial dye. Mater Res Express 4:55004. https://doi.org/10.1088/1758-5090/0/0/000000

Dąbrowski A, Hubicki Z, Podkościelny p, Robens E (2004) Selective removal of the heavy metal ions from waters and industrial wastewaters by ion-exchange method. Chemosphere 56:91–106. https://doi.org/10.1016/j.chemosphere.2004.03.006

Geny P, Dohen E (1972) Measures against water pollution in the iron and steel industry. Pure Appl Chem 29:191–200

Greenlee LF, Lawler DF, Freeman BD et al (2009) Reverse osmosis desalination: water sources, technology, and today's challenges. Water Res 43:2317–2348. https://doi.org/10.1016/j.watres.2009.03.010

Gupta A, Saurav JR, Bhattacharya S (2015) Solar light based degradation of organic pollutants using ZnO nanobrushes for water filtration. RSC Adv 5:71472–71481. https://doi.org/10.1039/C5RA10456D

Ince NH, Gönenç DT (1997) Treatability of a textile azo dye by UV/H2O2. Environ Technol 18:179–185. https://doi.org/10.1080/09593330.1997.9618484

Jin XC, Liu GQ, Xu ZH, Tao WY (2007) Decolorization of a dye industry effluent by *Aspergillus fumigatus* XC6. Appl Microbiol Biotechnol 74:239–243. https://doi.org/10.1007/s00253-006-0658-1

Kamdi MS, Khedikar IP, Shrivastava RR (2012) Physical and chemical parameter of effluent treatment plant for thermal power plant 1:1–10

Keck A, Klein J, Kudlich M et al (1997) Reduction of azo dyes by redox mediators originating in the naphthalenesulfonic acid degradation pathway of Sphingomonas sp. Reduction of Azo Dyes by Redox Mediators Originating in the Naphthalenesulfonic Acid Degradation Pathway of Sphingomonas sp. Str. Appl Environ Microbiol 63:3684–3690

Magic-Knezev A, van der Kooij D (2004) Optimisation and significance of ATP analysis for measuring active biomass in granular activated carbon filters used in water treatment. Water Res 38:3971–3979. https://doi.org/10.1016/j.watres.2004.06.017

Mathew Akinloye O, Olatunde Olubanjo F (2014) Assessment of the impacts of industrial effluent discharges on the water quality of Asa River, Ilorin, Nigeria. IOSR J Environ Sci Toxicol Food Technol 8:80–98. https://doi.org/10.9790/2402-08718098

Munter R (2003) Industrial wastewater characteristics. Balt Univ Program (BUP), Sweden 195–210

Peralta-Zamora P, Kunz A, Gomes de Moraes S et al (1999) Degradation of reactive dyes—I. A comparative study of ozonation, enzymatic and photochemical processes. Chemosphere 38:835–852. https://doi.org/10.1016/S0045-6535(98)00227-6

Pereira L (2012) Dyes—environmental impact and remediation. Springer, Netherlands, pp 111–162

Pokhrel D, Viraraghavan T (2004) Treatment of pulp and paper mill wastewater—a review. Sci Total Environ 333:37–58. https://doi.org/10.1016/j.scitotenv.2004.05.017

Robinson T, McMullan G, Marchant R, Nigam P (2001) Remediation of dyes in textile effluent: a critical review on current treatment technologies with a proposed alternative. Bioresour Technol 77:247–255. https://doi.org/10.1016/S0960-8524(00)00080-8

Zonoozi MH, Reza M, Moghaddam A, Arami M (2008) Removal of acid red 398 dye from aqueous solutions by coagulation/ flocculation process. Environ Eng Manag J 7:695–699

Chapter 11
Recent Advancement on Bioaugmentation Strategies for Process Industry Wastewater (PIWW) Treatment

Vivek Kumar, Madan Sonkar, Pooja and Sudheer Kumar Shukla

Abstract The quality of wastewater discharged from the process industries has high impact on all the biotic and abiotic components of the adjoining environment. Pollutants in the wastewater of different types of process industry include heavy metals, nitrates, sulfates, phosphates, fluorides, chlorides, oxalates, insecticides, herbicides, fungicides, polynuclear hydrocarbons (PAHs), phenols, polychlorinated biphenyls, halogenated aromatic hydrocarbons, formaldehyde, polybrominated biphenyls, biphenyls, detergents, oils, greases, hydrocarbons, alcohols, aldehydes, ketones, proteins, lignin. The presence of various pollutants in process industry wastewater (PIWW) directly or indirectly affects all life-forms in the vicinity. Since these are highly toxic, carcinogenic, mutagenic, and clastogenic, serious hazardous effects may run from aquatic organism to our food chain. Thus, PIWW treatment technologies are focused on three purposes, i.e., reduction in water usage, treatment of effluent, and recycling of water and other catalyst. In the last two decades, different types of physical–chemical treatment systems are established as secondary and tertiary stage wastewater treatment systems. However, these techniques are questionable due to the incurred cost, operational difficulties, and sludge generation. Augmentation of biological treatment seems to be a sustainable option in comparison with other physical and chemical processes. The biological treatment includes the aerobic and anaerobic digestion of PIWW. Ample amount of research has led to identify various aerobic and anaerobic bacterial strains having much higher treatment efficacy in comparison with the conventional activated sludge process. If these bacterial isolates are exploited in the form of consortium or

V. Kumar (✉)
Centre for Rural Development and Technology, Indian Institute of Technology, New Delhi 110016, Delhi, India
e-mail: vivekk@iitd.ac.in

M. Sonkar · Pooja
Department of Pulp and Paper Technology, Indian Institute of Technology, Roorkee Saharanpur Campus, Saharanpur 247001, India

S. K. Shukla
Department of Built and Natural Environment, Caledonian College of Engineering, Muscat, Oman

S. Bhattacharya et al. (eds.), *Water Remediation*, Energy, Environment, and Sustainability, https://doi.org/10.1007/978-981-10-7551-3_11

augmented into the activated sludge, they may exhibit even better results. Such aerobic and anaerobic bacterial treatment has been found to reduce the level of pollutant up to 90%. Additionally, sequential treatment PIWW can be attempted, but such trials are scarcely tested and are limited to the level of a pilot plant. In this chapter, emphasis is largely given on the various augmentation strategies considering various possible sequential treatments of PIWW. It is anticipated that the recent advancements in the area of bioaugmentation will certainly help in developing ecofriendly solutions to PIWW treatment.

11.1 Introduction

Water being a universal solvent and career of thermal energy is being utilized by various process industries (e.g., pulp and paper, petrochemical, textile, distillery, sewage, chemical, oil industries) in very large quantities (TERI Policy report 2015). Process industries also discharge large quantities of effluent. Due to the presence of various biorecalcitrant, toxic compounds, process industry wastewater (PIWW) has high impact on all the biotic and abiotic components of environment (Misra and Mani 1993; Saranya et al. 2014). These wastewaters in general characterized by parameters like pH, color, total dissolved solids (TDS), total organic carbon (TOC), total suspended solid (TSS), oil and greases, biochemical oxygen demand (BOD), chemical oxygen demand (COD), absorbable organic halides (AOX), extractable organic halides (EOX). They contain large number of compounds belonging to various groups like acrylon, alcohols, aldehydes, aniline, biphenyls, BTEX (benzene, toluene, ethylene, and xylene), calcium carbide, chlorides, chlorophenols (CHPs), detergents, dioxins, dye, fluorides, herbicides, formaldehyde, fungicides, glycol, halogenated aromatic hydrocarbons, salts of heavy metals, hydrocarbons, insecticides, ketones, lignin, naphthenic acids, nitrates, nitrobenzene, oils, oxalates, oxirene, pesticides, phenols, phosphates, phthalates, polyaromatic hydrocarbons (PAHs), polybrominated biphenyls, polychlorinated biphenyls, proteins, resin acids, sulfates, synthetic resin (Brito et al. 2015; Carey et al. 1991; Cunningham et al. 2001; Demeter et al. 2014; Haq et al. 2017; He et al. 2013; Liang et al. 2013; Lors et al. 2012; Ma et al. 2009; McKenzie et al. 2014; Ngah et al. 2011; Quero et al. 2015; Richardson 2013; Robinson et al. 2001; Tyagi et al. 2011; Veeresh et al. 2005). Broadly, pollutants coming out with PIWW can be classified as (i) organics (carbon-based compounds associated at one time or another with living things) and (ii) inorganics (dissolved metals, salts, etc.). Every waste stream has an inorganic component, but from treatment point of view, organic contents are of more significant for most wastewaters. The presence of this contamination on wastewater can bring about huge danger like variety of clastogenic, carcinogenic, endocrinic, mutagenic, and teratogenic effects in aquatic life and is running from aquatic organism to our food chain (Autenrieth et al. 1991; Savant et al. 2006). Some of the very common unfavorable impacts on flora and fauna are respiratory impact, oxidative anxiety, liver harm, and genotoxicity (Richardson et al. 2007).

During last decade, due to societal pressures, regulatory bodies in various developed and developing countries have emphasized on sustainability aspects of the river system through the strategy for zero discharge or treating effluent to a higher degree of purification, making them ecologically safe and then discharging into the river system. Conventional biological treatment options have poor efficiency toward biorefractories, large land area requirement, less flexibility in design and operation, sensitivity toward diurnal variation, toxicity of some chemicals and environmental factors (Bhattacharya et al. 2003). Therefore, an extensive treatment methodology is necessary for removal or demineralization of pollutants present in the complex wastewaters of process industries to meet the highest safety standards. Different types of physical–chemical treatment systems have been established for achieving high level of purification of PIWW. These physicochemical processes are mainly separation processes in which organics are not mineralized but rather change their state and generate a low-volume highly concentrated polluted stream, the treatment of which always poses a major challenge. These systems are based mainly on the principal of precipitation, coagulation, adsorption (Mohan et al. 2004), oxidation (Padoley et al. 2011), electrocoagulation (Soloman et al. 2009), ultrasound (Shaw and Lee 2009), reverse osmosis (Chang et al. 2016), and photocatalytic oxidation (Jamil et al. 2011). Additionally, physicochemical treatment strategy produces aromatic compounds, free radicals, sludge, etc., which further poses environmental problems (Robinson et al. 2001).

In another approach, catalytic oxidation is used for pretreatment of wastewaters. In pretreatment, pollutant chemically transforms complex form to simple compound, and then, treated effluent is subjected to biological treatment which mineralize the simple compound. Thus, catalytic oxidation is used to enhance the efficiency of the next stage. Catalytic oxidation is principally based on different catalysts, viz. titanium dioxide (TiO_2), zinc oxide (ZnO), ultraviolet (UV)/photoradiation, hydrogen peroxide, Fenton's reagent (H_2O_2/Fe^{2+}), UV, UV/H_2O_2, photo-Fenton (UV/H_2O_2/Fe^{2+}), ozonation, peroxone (ozone/H_2O_2). (Eskelinen et al. 2010). However, none of these advancements have gained wide acceptance in process industries (Mojiri et al. 2013).

Since last two and half decades, researchers are working to improve the efficacy of biological treatment systems of wastewaters and several biological techniques, such as bioremediation and phytoremediation, have been developed, so that biorecalcitrant and toxic compounds can be mineralized or converted into simpler and non-toxic compounds without facing challenges of physical–chemical treatments (Schoefs et al. 2004). Presently, bioremediation is the most environmentally benign and economical methods of remediation of contaminated environment including wastewaters (Zhang and Quiao 2002). Bioremediation increases the natural microbial degradation rate of pollutants by supplementing the microbial culture (bacteria, fungi, or consortium) with substrates such as carbon sources or electron donors (biostimulation, biorestoration). Another strategy of bioremediation is enriching of microbial culture with a specific microorganism that has pathways to degrade the targeted pollutant at higher rate (Mackay and Fraser 2000). The aim of bioremediation is to reduce the levels of targeted pollutants to environmentally ecologically safe or ideally, to completely demineralize pollutants.

By employing a specific microorganism adapted to the wastewater, bioaugmentation technologies can effectively enhance the removal efficiency of refractory organics. Bioaugmentation includes utilization of enzymes, indigenous or engineered strains of bacteria, fungi, and algae (Bajpai and Bajpai 1994), microbial consortia (Pendashteh et al. 2010), and additionally, a combination of bacterial–fungal treatment capabilities to partially degrade or completely mineralize the targeted compounds.

However, identification of individual strains or microbial consortia that possess unique metabolic capabilities required to biodegrade or biotransform target pollutants is the primary requirement of the successful implementation of any bioaugmentation strategy.

11.2 Augmentation of the Biological Systems to Improve the Process Industry Wastewater Treatment Efficiency

A novel approach that has received attention in recent time is bioaugmentation which works on biorefractories through metabolic or catabolic pathways carrying specialized bacterial, fungal enzymes and mixed culture of fungus and bacteria (Hai et al. 2012).

In bioaugmentation strategy, specific compounds are targeted for which individual strains or microbial consortia are identified and isolated. These individual strains or consortia degrade the targeted compound through metabolic or catabolic pathway and either biodegrade or biotransform target pollutants. Biologically mediated reactions are catalytic processes of two different types: (i) metabolic and (ii) cometabolic. Metabolic reactions are growth-linked processes that often result in mineralization of the micropollutant, while cometabolic reactions are the processes which do not sustain growth of the responsible microorganisms and often lead to the generation of transformed products (Tran et al. 2013). A schematic diagram of the metabolic and cometabolic strategies is shown in (Fig. 11.1).

11.2.1 Enzymatic Treatment

The treatment of PIWW through enzymes which have the potential to degrade the pollutants and organics present in effluent has a high potential for field application. There are different types of enzymes known for the degradation of various contaminants such as manganese peroxidase (MnP), lignin peroxidase (LiP) (Wesenberg et al. 2003), laccase (Hadibarata and Kristanti 2014), horseradish peroxidase (HRP) (Forgiarini and de Souza 2007) etc.

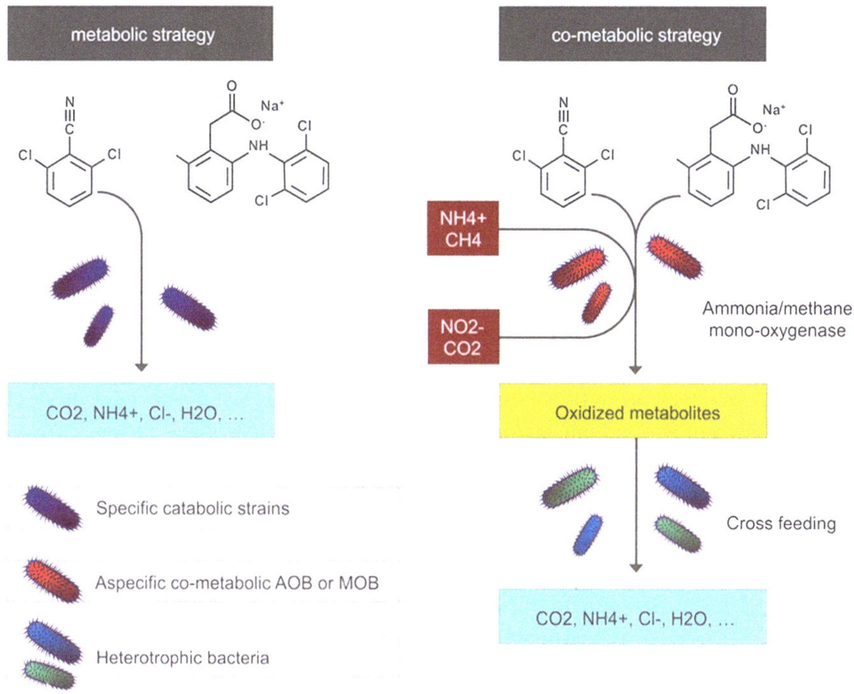

Fig. 11.1 Metbolic and cometabolic strategies

Some of the enzymes may decolorize and detoxify the organic contaminants present in PIWW industry effluents. The utilization of microbial or enzymatic treatment offers benefits over physical and/or chemical decolorization and AOX precipitation techniques. However, biochemical instability and issues pertaining to the reusability of the enzyme are some of the disadvantages of enzymatic processes. Immobilization of the enzyme is necessary for the stability of biochemical strength and reusability of the enzymes.

11.2.2 Sequential Aerobic and Anaerobic Microbial Treatment of Pollutants

Decomposition of pollutants present in the PIWW depends on their bonding nature and arrangement of molecules. In aerobic condition (presence of oxygen), organics are oxidized, whereas in the anaerobic condition (absence of oxygen), most of the pollutants are reduced. For the mineralization and conversion, value-added products of organic waste generated from PIWW aerobic–anaerobic process can be used. Degradation pollutants take place through microbes separately in two stages: reductive cleavage and aerobic degradation.

Robinson et al. (2001) indicated that under aerobic condition, azo dyes are not readily metabolized. However, different types of dye molecules which have electrophilic azo bond are reduced (breaking of the 'azo linkage') by many bacteria in anaerobic condition to produce colorless aromatic amines, and then, aerobic degradation for the removal can be followed (Gomes et al. 2000). More than 20 anaerobic bacterial strains have been isolated and detected (16s rRNA) from wastewater sludge (Sekiguchi et al. 2001; Sekiguchi and Kamagata 2004; Sekiguchi 2006); out of these, maximum are Methanogenens i.e., Proteobacteria, Chloroflexi, Firmicutes, Spirochetes, and Bacteroidetes (Sekiguchi and Kamagata 2004). Therefore, a wastewater containing azo dyes can be treated successfully by employing anaerobic and aerobic processes in combination. The mineralization of azo dyes requires integrated or sequential anaerobic and aerobic stages. In anaerobic step, azo dyes are biotransformed, and in aerobic step, transformed products are biodegraded. Sequential microaerophilic/aerobic reactors using *Staphylococcus arlettae* are reported in the literature for biodegradation of various other dyes (Elisangela et al. 2009; Moutaouakkil et al. 2004).

11.2.3 Bioaugmentation Strategies for Various Process Industry Wastewaters

11.2.3.1 Pulp and Paper Industries

The pollutants in pulp and paper manufacturing plant are generated mostly by two processes, i.e., pulping and bleaching. Bleaching effluent is heavily loaded with organic compounds, especially AOX, and needs specialized treatment. Besides, effluent from pulp and paper industries, in particular that from the industries which produce writing and printing grade of pulp and paper, has low biodegradability due to the high amount of CHPs (Pokhrel and Viraghavan 2004). It contains many different types of organic contaminants, like CHPs, resins and fatty acids, dioxins, and furans, which are not quantifiable through BOD. These contaminants result in color and toxicity of the effluent. Researchers have reported more than 500 organic compounds in the effluents of bleach plants. Out of these 500 organic compounds, approximately 330 compounds include acids, esters, alcohols, resin fatty acids, dioxins, furans and 'chlorinated organics' such as chlorate, chloroform, chlorophenols, chlorocatechols, chloroguaiacols, chlorosyringols, chlorosyringaldehydes and chlorovanillins. (Mckague and Carlberg 1996).

Among the chlorinated phenolics, tri-chloroguaiacols and tetra-chloroguaiacols are most widely reported and identified in the bleached kraft and pulp industry effluents (Liebergott et al. 1990). Chlorinated dioxins are present in very low concentrations (generally in ppm levels), of the aggregate AOX released with the bleach plant effluent. Around 210 dioxins, belonging to two families, polychlorinated dibenzo-*p*-dioxins (PCDDs) and polychlorinated dibenzo-*p*-furans (PCDFs) have been accounted in the bleach effluents. 2, 3, 7, 8-tetrachlorodibenzo-*p*-furans

(2, 3, 7, 8-TCDF) and 2, 3, 7, 8-tetrachlorodibenzo-*p*-dioxin (2, 3, 7, 8-TCDD) are critical from the toxicological perspective owing to its highly carcinogenic and biomagnification properties (Carey et al. 1991).

Indigenous Microorganisms for the Treatment of Pulp and Paper Mill Effluent

Indigenous microorganisms present in the polluted environment possess great potential to transform and/or degrade recalcitrant compound. Due to the continuous exposure to a long duration, microbes in the environment like water streams, soils (Sorensen et al. 2007) and wastewater treatment plant (Bouju et al. 2012) develop not only resistance but mineralizing capabilities against specific micropollutants. These indigenous microbes adapt to the environmental conditions, and their genetic structures modify accordingly through horizontal gene transfer, mutation, and genetic rearrangements (Boon et al. 2000). There are good numbers of bacterial consortium (*Pseudomonas putida*, *Citrobacter* sp., *Enterobacter* sp./*Klebsiella* sp., *Alcaligens* sp., *Cronobacter* sp./*Pseudomonas*, *Bacillus*, *Pannonibacter*, *Ochrobacterum*/ *Pseudomonas aeruginosa* (DSMZ 03504), *Pseudomonas aeruginosa* (DSMZ 03505), *Bacillus megaterium* (MTCC 6544)/*Pseudomonas aeruginosa* (PAO1), *Stenotrophomonas maltophila, Proteus mirabilis*/*Paenibacillus* sp., Aneurinibacillus aneurinilyticus, Bacillus sp./Bacillus sp., Serratia marcescens/*Flavobacterium gleum*, *Agrobacterium radiobacter*, *Pseudomonas* sp.) reported in the literature, which has been utilized to decrease the organic load up to 80–90% from the pulp and

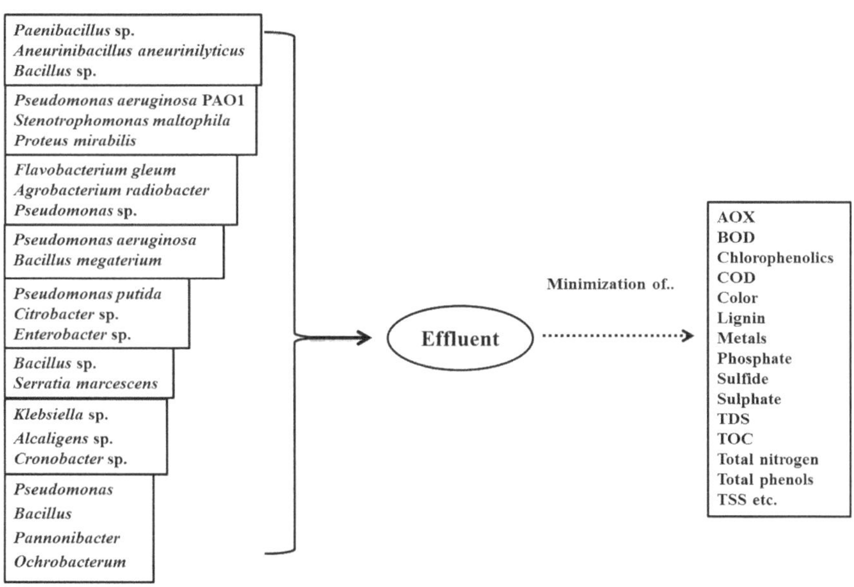

Fig. 11.2 Reduction of pollutant in PAP industries through bacterial consortium

paper industry effluents (Chandra and Abhishek 2011; Chandra and Singh 2012; Fulthorpe and Allen 1995; Kumar et al. 2012, 2014; Paliwal et al. 2015; Sharma and Thakur 2009; Sharma et al. 2010; Singh et al. 2008; Srivastava and Thakur 2007; Tiku et al. 2010). They are not only effective in reducing pollution load in terms of effectiveness for the degradation of COD, BOD, Color, AOX, and TDS but also reduce microorganic pollutants like chlorophenol, lignin (Fig. 11.2).

There are limited fungal consortiums, which have shown potential for the treatment of pulp and paper industry effluent. Two basidiomycetous fungi (*Merulius aureus* and an unidentified variety) and a *Fusarium sambucinum* Fuckel MTCC 3788 were identified from soils contaminated by pulp and paper industry effluents over a long time. Further, the consortia of the above isolated fungal strains were utilized for the treatment of PAP industry effluents (Malaviya and Rathore 2007).

Sequential Treatment of Pulp and Paper Industry Effluents

Sequential bioaugmentation strategy means two or more microbial strains can be used sequentially (Chuphal et al. 2005; Singh and Thakur 2006) and give result for more than 90% degradation of pollutants. In this strategy, first strain will be involved in the catabolic reaction and transform the biorecalcitrant compounds into simpler compounds, which gets mineralized completely in the subsequent stage through the metabolic pathways. Mineralization of bicyclic aromatic compounds like chlorinated biphenyls, chlorinated di-benzofurans, and naphthalene sulfonates, are reported in the literature using the catabolic pathways (Pokhrel and Viraraghavan 2004; Singh 2007). It is imperative that sequential treatment can certainly have major advantages over single-stage process in terms of alleviating multiple parameters at a time. However, very few literature is available on the exploitation of microbial isolates in sequential treatment of pulp and paper industry effluents.

11.2.3.2 Petrochemical Industries

Petrochemical industries have gain utmost importance in recent times due to increasing use of petroleum-based products for the industrial development throughout and so the concern regarding the petrochemical industrial effluent. Petrochemical effluents have greatly affected the surface water systems. It has been established that petrochemical wastewaters cause the release of genotoxins, into the environment and badly affect the aquatic environment (White et al. 1996). The volume and characteristics of petrochemical effluents largely depend on the process configuration. However, all process results into large amount of organic refractories in wastewaters. The petrochemical refractory organics including hydrocarbon, aniline, nitrobenzene, organochlorines, phenols, VOC are most frequently present in the effluents. VOCs are volatile compounds and escape to the atmosphere creating sometimes odorous and otherwise hazardous environment resulting into noxious, mutagenic, and carcinogenic effects. Different types of methods (carbon adsorption, biofiltration, incineration) are reported in the literature for the removal and control of VOC (Guo et al. 2009).

Conventional activated sludge methods with pretreatment for oil/water separation have been modified for the treatment of effluent in petrochemical but not resulting into the significant improvement in the removal efficiency of the pollutant because of the presence of phenolic, phthalic compounds, aromatic amines, etc., which cause hampering of the bacterial growth and adversely affect the operational facility.

Complexity of petrochemical wastewaters makes wastewater treatment plants (WWTPs) in the petrochemical industry often unstable. The operation performance of WWTPs is further affected due to frequent shock loading especially under the conditions with low temperature at which the growth, reproduction, and metabolism of the microorganisms are further inhibited (Nyhoim 1996; Patel and Madamwar 2002; Zhao et al. 2009).

Bioaugmentation strategies of mixing specialized bacteria exposed petrochemical wastewater for long time in the microbial culture proved to be very effective in improving the removal efficiency of refractory organics (Hu et al. 2008; Kasai et al. 2007; Margesin and Schinner 1999). Bioaugmentation strategies were implemented quite successfully in laboratory-scale activated sludge and other bioreactors (Boon et al. 2003; Maes et al. 2006, Mohan et al. 2005; Quan et al. 2004); however, full-scale application of bioaugmentation reported rarely in the literature may be due to the complex composition of wastewater and the practical operational difficulties of bioaugmentation strategies. In the present study, long-run experiments will be carried out on a laboratory-scale experimental unit. Long-run data shall be used to simulate and design the field plant.

The success of biological treatment mainly depends on bioaugmentation and behavior of the inoculated strains (more resistant to polluted environment) where it takes place. Therefore, it maintains the population of microbial strain into the environment in spite of some microbial inoculums dead, wash out or predation by other microbes like protozoa in the treatment plant (El Fantroussi and Agathos 2005). It is also noticeable that the periodic addition of isolated strain could provide the system with specific sufficient biomass, and it could not justify the complex operation and high cost. For the proper activity of inoculated strains without influences, other microbes establish the bioaugmentation upgrading contact oxidation process. It proved to be a good solution toward the prevention of the microorganisms from being washed out or grazed by other microorganisms such as protozoa (Daane and Häggblom 1999; El Fantroussi and Agathos 2005). Zhan et al. (2006) worked on the contact oxidation tank, bioaugmentation with mixed culture of specialized bacteria targeting to various refractory organics. The mixed culture that includes bacterial strains *Pseudomonas*, *Bacillus*, *Acinetobacter*, *Flavobacterium,* and *Micrococcus* had been enriched from the activated sludge of various petrochemical wastewaters through isolation and acclimation. By opting this method, the organic loading rate (OLR) was increased stepwise from 0.04 to 0.5 kg COD/m^3 d at the end of the upgrading period as the flow rate reached the design value of 700 m^3/d.

Combined bioaugmented and membrane separation-based systems are becoming popular and exist at several petrochemical units in the pilot plant stage. However, due to the presence of contaminants into the wastewater, extensive pretreatment is often required to maintain high performance (Ravanchi et al. 2009) of such system.

11.2.3.3 Textile Industries

Textile industries generate and discharge various types of pollutants into the environment. Textile effluent is very complex in nature and contains highly polluting compounds including dyes (Robinson et al. 2001). Apart from other dyes, azo dyes are highly consumed (60–70%) in textile processing industries (Hunger et al. 2004). Majority of dyes present in textile effluent including azo dyes are xenobiotics, and they tend to persist under aerobic environment because of their electron-withdrawing nature (Knackmuss 1996). Reductive cleavage of azo bond (–NN–) has been reported under anaerobic condition and decolorizes the textile effluent (Van der Zee and Villaverde 2005). The reduction of many azo dyes is, however, a rather slow process (Kapdan et al. 2003; Méndez-Paz et al. 2005).

It is very important to know that unlike bacterial activated sludge, aerobic white-rot fungi have potential to degrade varieties of recalcitrant compound and different type of textile dyes (Fu and Viraraghavan 2001). Previous works done by different researchers have shown that fungi have great potential to remove the color and total organic carbon of effluent from textile industries (Hai et al. 2008; Kaushik and Thakur 2009; Sahoo and Gupta 2005; Singhal and Thakur 2009). However, textile wastewater after anaerobic treatment carries many other colorless organics which are complex and difficult to degrade like aromatic amines. Many of these organics are very toxic and generate during the anaerobic reduction of azo dyes (O'Neill et al. 2000). Although some researchers have reported aerobic removal of aromatic amines (IsIk and Sponza 2004), it still remains debatable (Lourenco et al. 2000). Therefore, decolorization is not the only performance parameter of WWTPs in textile processing unit, but also TOC removal is very critical to monitor.

Degradation of xenobiotics, organics, and other dyes has been widely reported in the literature using bioaugmented systems working on mixed microbial cultures. Several researchers have reported that a higher level of biodegradation and mineralization can be achieved through cometabolic activities which take place inside a microbial community complementing each other. The catabolic pathways of fungal enzyme-like laccase have likewise been found responsible for dye decolorization and degradation of azo dyes from textile processing effluent (Chivukula and Renganathan 1995; Trejo-Hernandez et al. 2001). The metabolic pathways of mixed cultures of *Trametes versicolor* and *Pseudomonas putida* have been reportedly used for opening of the Acid Red 27 dye structure and subsequent degradation (Hai et al. 2012). Knackmuss (1996) reported the degradation of naphthalene sulfonates by a two-species mixed culture. *Sphingomonas* strain BN6 decomposes naphthalene-2-sulfonate into salicylate ion equivalents which are actually toxic to strain BN6; therefore, in two-spices mixed culture, an organism which is capable of degrading the salicylate ion is used as second complementary specie. Isolation of a single bacterial strain from dye-containing wastewater samples is a very difficult task, and quite often, long-term adaptation strategies are employed before the isolate is capable of using the pollutants as a respiratory substrate. Several works have been reported in the literature, related to the biodegradation ofcolored textile wastewater using mixed bacterial cultures.

Hai et al. (2011) reported almost complete removal of TOC using granular activated carbon (GAC)-packed aerobic reactor. The GAC played the key role in decoloration, while the aerobic zone was vital for TOC removal. Hai et al. (2008) designed a membrane bioreactor containing a mixed microbial culture dominating with fungi. The reactor reported a higher removal of color and TOC in comparison with the conventional MBR.

11.2.3.4 Distillery Industries

Like other industries, distillery industry uptakes large volumes of water and discharges highly polluted wastewater that is of serious environmental concern. The distillery wastewater is characterized by highly loaded chemical oxygen demand (COD) (80,000–100,000 mg/l) and biochemical oxygen demand (BOD) (40,000–50,000 mg/l), apart from low pH, strong odor, and dark brown color (CPCB 2003). The generation of wastewater from different stages of the distillation process in distillery industry is termed 'spent wash.' Spent wash formed the combination of amino acids and reducing sugars at high temperatures which is very harmful colored compounds (Satyawali and Balakrishanan 2008). Alcohol manufacturing in distilleries follows four main steps, viz. feed preparation, fermentation, distillation, and packaging (Fig. 11.3) (Satyawali and Balakrishanan 2008). Various researchers investigated and reported the presence of various genotoxic compounds in the wastewaters discharged from distilleries (Mahimaraja 2004; Chandra et al. 2008) regularly. Thus, it is very necessary to take proper remedial measures before final disposal of distillery effluents.

Kaushik et al. (2010) isolated microbial strains (fungi and bacteria) from the distillery mill site and used them for the removal of different nature of pollutants from distillery effluent. Strains were used in two stages in sequential manner. In first stage, fungal strain was used which could effectively absorb the shock loads, and in second stage, bacterial strain was used. Fungal strain resulted in stable performance besides the removal of colored compounds, whereas bacterial strain was effective in removing most of the recalcitrant organic compounds and drastically reducing the COD levels of the distillery wastewater.

White-rot fungus secretes ligninolytic enzymes, and these enzymes are reported to degrade xenobiotics and organo-compounds found in distillery effluents. Among the fungal strains, *Phanerochaete chrysosporium* and *Trametes versicolor* are most widely studied among these (Gonzalez et al. 2000). *P. chrysosporium* JAG 40 was used for decolorization of diluted synthetic melanoidin as well as for treating 6.25% anaerobically digested spent wash and resulted in 80% removal of color (Dahiya et al. 2001; Kumar et al. 1998). The fungus removes 82% color when used for the treatment of 12.5% anaerobically–aerobically treated effluent. Out of the total color

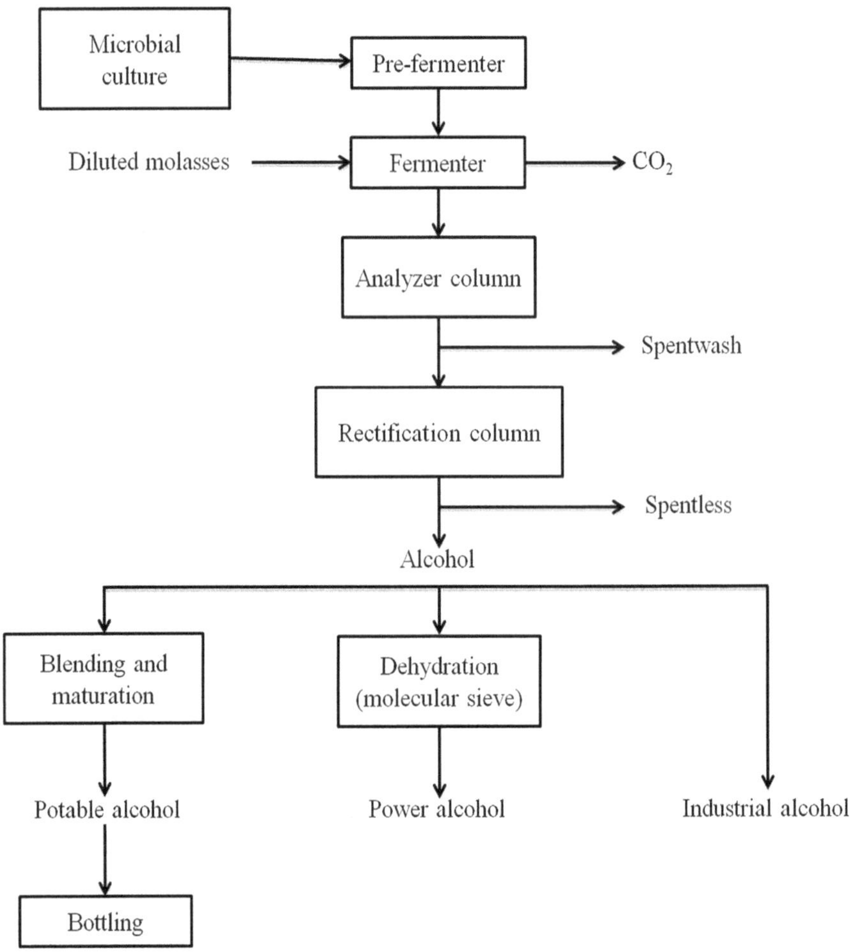

Fig. 11.3 Process description: feed preparation, fermentation, distillation, and packaging

removal, 90% is reported to be removed biologically and the rest through adsorption on the mycelium (Dehorter and Blondeau 1993). Miyata et al. (1998) reported the decolorization of melanoidin pigment using extracellular H_2O_2 and peroxidase which is produced by fungus *Coriolus hirsutus*. Kumar et al. (1998)also used the fungus *C. hirsutus* for the treatment of anaerobically digested spent wash and successfully reduced color by 71–75% and COD by 90%. Chopra et al. (2004) used *C. versicolor* for the treatment of biodigested distillery wastewater. Friedrich (2004) found *F. flavus* capable of detoxifying the distillery effluent due to (68%) reduction in PAH. Besides, it also results (73%) in color removal.

Aspergillus sp is among the most popular filamentous species which has been reported to be used for the treatment of distillery wastewaters (Friedrich, 2004). *A. niveus* and *A. niger* were found capable of reducing color up to 60–69% and COD up to 75–95%. Enhancement of seedling growth in Zea mays using treated effluent was also reported in the literature (Angayarkanni et al. 2003; Miranda et al. 1996). Ghosh et al. (2002) studied the use of two-stage bioreactor for treating the distillery wastewaters. First-stage reactor based on *Pseudomonas putida* was followed by *Aeromonas* sp.-based second stage reactor. The two-stage bioreactor significantly reduces COD as well as color. One of the mechanisms satisfied for color removed by *P. putida* produces hydrogen peroxide which is a strong decolorizing agent. Besides (60%) color reduction, *P. putida* also result in (40%) COD reduction. Jain et al. (2001, 2002) carried out the treatment of predigested distillery effluent with *Aeromonas formicans*, and observed (57%) COD reduction and (55%) decease in color during the study, also by the using *B. cereus* the reduction of COD and color was (81%) and (75%), respectively. Kumar and Chandra (2006) use three isolated strains *Bacillus thuringiensis*, *Bacillus brevis,* and *Bacillus* sp. for treating distillery wastewater. The level of decolorization of melanoidins by each isolated was observed between 1–31% range, however when used in consortium these isolates

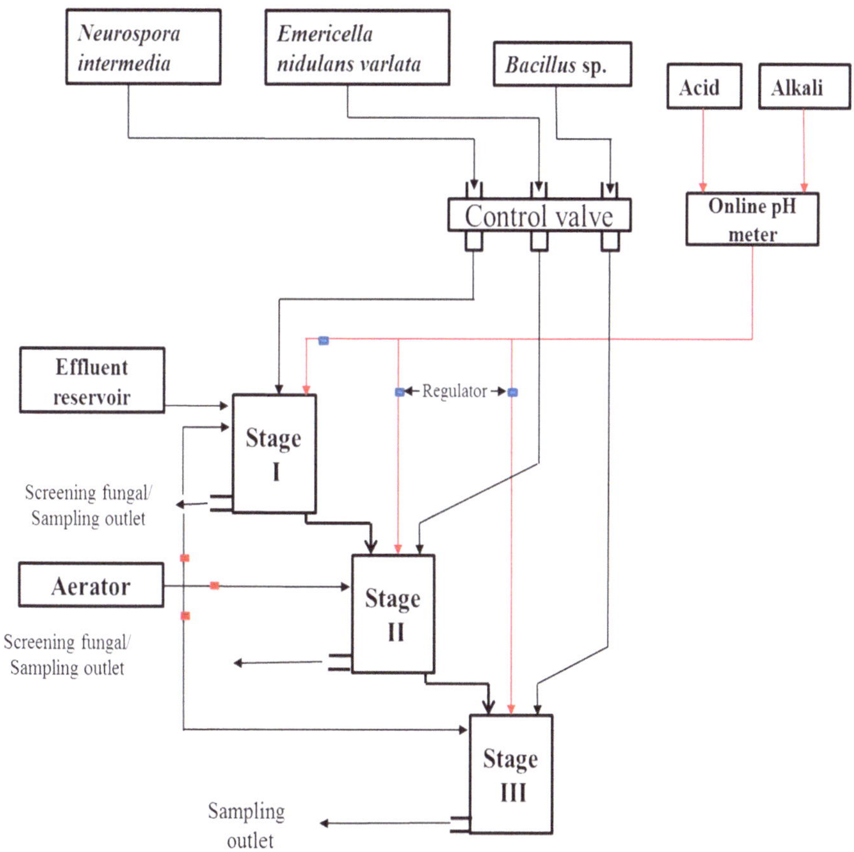

Fig. 11.4 Schematic diagram of three-stage bioreactor for the treatment of distillery spent wash

result up to 50% decolorization, and that is due to the enhanced effect of coordinated metabolic interactions.
Multistage sequential biological treatment of effluent has been used very effectively by Kaushik et al. (2010) in achieving very high treatment efficiencies, and the treatment was carried out in three stages with 10% (v/v) MSM effluent. Treatment was done by type I fungus in stage I followed by type II fungus in stage II and finally in stage III by bacteria. A schematic diagram of three-stage bioreactor is shown in (Fig. 11.4).

The treated effluent showed significant reduction in color (82%) and COD (93%). The data imply that stepwise treatment using fungi and bacteria yielded faster and better results in comparison with individual strain treatments.

11.3 Conclusion

Process industries consume high amount of water in their production process and also generate high amount of wastewaters of complex nature. The contamination present in wastewaters can bring about huge danger to flora and fauna such as aquatic life and running them from aquatic organism to our food chain. It contains so many different types of refractory organic contaminants which are not quantifiable through BOD and hence goes unreported and remains unnoticed. These contaminants generally result in color and toxicity in terms of clastogenic, carcinogenic, endocrinic, and mutagenic effects.

In order to ensure the production of high-quality treated wastewater from process industries, bioaugmented treatment systems may be regarded as the most attractive among the known techniques. By employing a specific microorganism adapted to the wastewater, bioaugmentation technology can effectively enhance the removal of refractory organics and toxic compounds. Bioaugmentation strategies include utilization of enzymes, fungus, and bacteria as single strains, microbial consortium, and additionally combination of bacterial–fungal treatment that can be able to partially degrade or mineralize target compounds. Microbes can be isolated from the site where they survive after long time of exposure to effluent.

Bacterial consortium is the better option for reducing the pollution load as compared to single isolates. Although fungal strains are the good pollutant degraders, they are not easy to handle. There are limited fungal consortium known for the treatment of paper mill effluent, and it is because of their resistance activity and larger incubation period. Also, sequential treatment could be a strong approach for the treatment of PIWW effluents.

References

Angayarkanni J, Palaniswamy M. Swaminathan K (2003) Biotreatment of distillery effluent using *Aspergillus niveus*. Bull Environ Contam Toxicol 70(2):0268–0277. doi:https://doi.org/10.1007/s00128-002-0187-2

Autenrieth RL, Bonner JS, Akgerman A, Okaygun M, McCreary EM (1991) Biodegradation of phenolic wastes. J Hazard Mater 28:29–53. https://doi.org/10.1016/0304-3894(91)87004-L

Bajpai P, Bajpai PK (1994) Biological colour removal of pulp and paper mill wastewaters. J Biotechnol 33(3):211–220. https://doi.org/10.1016/0168-1656(94)90069-8

Bhattacharya S, Sharma A, Nayak SK, Chattopadhyay S, Mukherjee AK (2003) NMR study of complexation of crown ethers with [60]-and [70] fullerenes. J Phys Chem B 107(18):4213–4217. https://doi.org/10.1021/jp022157e

Boon N, Goris J, De VP, Verstraete W, Top EM (2000) Bioaugmentation of activated sludge by an indigenous 3-chloroaniline—degrading *Comamonas testosteroni* strain I2gfp. Appl Environ Microbiol 66(7):2906–2913. https://doi.org/10.1128/AEM.66.7.2906-2913.2000

Boon N, Top EM, Verstraete W, Siciliano SD (2003) Bioaugmentation as a tool to protect the structure and function of an activated-sludge microbial community against a 3-chloroaniline shock load. Appl Environ Microbiol 69(3):1511–1520. https://doi.org/10.1128/AEM.69.3.1511-1520.2003

Bouju H, Ricken B, Beffa T, Corvini PFX, Kolvenbach BA (2012) Isolation of bacterial strains capable of sulfamethoxazole mineralization from an acclimated membrane bioreactor. Appl Environ Microbiol 78(1):277–279. https://doi.org/10.1128/AEM.05888-11

Brito EM, De la Cruz' Barrón M, Caretta CA, Goñi-Urriza M, Andrade LH, Cuevas-Rodríguez G, Malm O, Torres JP, Simon M, Guyoneaud R (2015) Impact of hydrocarbons, PCBs and heavy metals on bacterial communities in Lerma river, Salamanca, Mexico: investigation of hydrocarbon degradation potential. Sci Total Environ 521:1–10. doi:https://doi.org/10.1016/j.scitotenv.2015.02.098

Carey PJ, Oberst MT, McCubbin MA, Hughes SH (1991) Appraisal and caregiving burden in family members caring for patients receiving chemotherapy. Oncol Nurs Forum 8(8):1341–1348

Central Pollution Control Board (CPCB) (2003) Environmental management in selected industrial sectors status and needs, PROBES/97/2002–03, CPCB. Ministry of Environment and Forest, New Delhi, India

Chandra R, Abhishek A (2011) Bacterial decolorization of black liquor in axenic and mixed condition and characterization of metabolites. Biodegradation 22(3):603–611. https://doi.org/10.1007/s10532-010-9433-1

Chandra R, Singh R (2012) Decolourisation and detoxification of rayon grade pulp paper mill effluent by mixed bacterial culture isolated from pulp paper mill effluent polluted site. Biochem Eng J 61:49–58. https://doi.org/10.1016/j.bej.2011.12.004

Chandra R, Bharagava RN, Rai V (2008) Melanoidins as major colourant in sugarcane molasses based distillery effluent and its degradation. Biores Technol 99(11):4648–4660. https://doi.org/10.1016/j.biortech.2007.09.057

Chang H, Liang H, Qu F, Ma J, Ren N, Li G (2016) Towards a better hydraulic cleaning strategy for ultrafiltration membrane fouling by humic acid: effect of backwash water composition. J Environ Sci 43:177–186. https://doi.org/10.1016/j.jes.2015.09.005

Chivukula M, Renganathan V (1995) Phenolic azo dye oxidation by laccase from Pyriculariaoryzae. Appl Environ Microbiol 61(12):4374–4377

Chopra P, Singh D, Verma V, Puniya AK (2004) Bioremediation of melanoidin containing digested spentwash from cane-molasses distillery with white rot fungus Coriolusversicolor. Indian J Microbiol 44:197–200

Chuphal Y, Kumar V, Thakur IS (2005) Biodegradation and decolorization of pulp and paper mill effluent by anaerobic and aerobic microorganisms in a sequential bioreactor. World J Microbiol Biotechnol 21(8):1439–1445. https://doi.org/10.1007/s11274-005-6562-5

Cunningham JA, Rahme H, Hopkins GD, Lebron C, Reinhard M (2001) Enhanced in situ bioremediation of BTEX-contaminated groundwater by combined injection of nitrate and sulfate. Environ Sci Technol 35(8):1663–1670. https://doi.org/10.1021/es001722t

Daane LL, Häggblom MM (1999) Earthworm egg capsules as vectors for the environmental introduction of biodegradative bacteria. Appl Environ Microbiol 65(6):2376–2381

Dahiya J, Singh D, Nigam P (2001) Decolourisation of molasses wastewater by cells of Pseudomonas fluorescens immobilised on porous cellulose carrier. Biores Technol 78(1):111–114. https://doi.org/10.1016/S0960-8524(00)00163-2

Dehorter B, Blondeau R (1993) Isolation of an extracellular Mn-dependent enzyme mineralizing melanoidins from the white rot fungus *Trametes versicolor*. FEMS Microbiol Lett 109(1):117–122

Demeter MA, Lemire J, George I, Yue G, Ceri H, Turner RJ (2014) Harnessing oil sands microbial communities for use in ex situ naphthenic acid bioremediation. Chemosphere 97:78–85. https://doi.org/10.1016/j.chemosphere.2013.11.016

El Fantroussi S, Agathos SN (2005) Is bioaugmentation a feasible strategy for pollutant removal and site remediation? Curr Opin Microbiol 8(3):268–275. https://doi.org/10.1016/j.mib.2005.04.011

Elisangela F, Andrea Z, Fabio DG, de Menezes Cristiano R, Regina DL, Artur CP (2009) Biodegradation of textile azo dyes by a facultative *Staphylococcus arlettae* strain VN-11 using a sequential microaerophilic/aerobic process. Int Biodeterior Biodeg 63(3):280–288. https://doi.org/10.1016/j.ibiod.2008.10.003

Eskelinen K, Sarkka H, Kurniawan TA, Sillanpaa MET (2010) Removal of recalcitrant contaminants from bleaching effluents in pulp and paper industries using ultrasonic irradiation and Fenton-like oxidation, electrochemical treatment, and/ or chemical precipitation: a comparative study. Desalination 255(1–3):179–187. https://doi.org/10.1016/j.desal.2009.12.024

Forgiarini E, de Souza AAU (2007) Toxicity of textile dyes and their degradation by the enzyme horseradish peroxidase (HRP). J Hazard Mater 147(3):1073–1078. https://doi.org/10.1016/j.jhazmat.2007.06.003

Friedrich J (2004) Bioconversion of distillery waste. In: Fungal biotechnology and agricultural, food, and environmental applications. CRC Press, Boca Raton, USA, pp. 431–442

Fu Y, Viraraghavan T (2001) Fungal decolorization of dye wastewaters: a review. Biores Technol 79(3):251–262. https://doi.org/10.1016/S0960-8524(01)00028-1

Fulthorpe RR, Allen DG (1995) A comparison of organochlorine removal from bleached kraft pulp and paper –mill effluents by dehalogenating *Pseudomonas*, *Ancylobacter* and *Methylobacterium* strains. Appl Microbiol Biotechnol 42(5):782–789. https://doi.org/10.1007/BF00171962

Ghosh M, Ganguli A, Tripathi AK (2002) Treatment of anaerobically digested distillery spentwash in a two-stage bioreactor using *Pseudomonas putida* and *Aeromonas* sp. Process Biochem 37 (8):857–862. https://doi.org/10.1016/S0032-9592(01)00281-3

Gomes A, Brás R, Ferra MIA, Amorim MTP, Porter RS (2000) Biological treatment of effluent containing textile dyes. Color Technol 116(12):393–397. https://doi.org/10.1111/j.1478-4408.2000.tb00016.x

González T, Terrón MC, Yagüe S, Zapico E, Galletti GC, González AE (2000) Pyrolysis/gas chromatography/mass spectrometry monitoring of fungal-biotreated distillery wastewater using *Trametes sp.* I-62 (CECT 20197). Rapid Commun Mass Spectrom 14(15):1417–1424. https://doi.org/10.1002/1097-0231(20000815)14:15

Guo J, Ma F, Chang CC, Cui D, Wang L, Yang J, Wang L (2009) Start-up of a two-stage bioaugmented anoxic–oxic (A/O) biofilm process treating petrochemical wastewater under different DO concentrations. Biores Technol 100(14):3483–3488. https://doi.org/10.1016/j.biortech.2009.02.059

Hadibarata T, Kristanti RA (2014) Potential of a white-rot fungus *Pleurotus eryngii* F032 for degradation and transformation of fluorene. Fungal Biol. 118(2):222–227. https://doi.org/10.1016/j.funbio.2013.11.013

Hai FI, Yamamoto K, Nakajima F, Fukushi K (2008) Removal of structurally different dyes in submerged membrane fungi reactor—biosorption/PAC-adsorption, membrane retention and biodegradation. J Membr Sci 325(1):395–403. https://doi.org/10.1016/j.memsci.2008.08.006

Hai FI, Yamamoto K, Nakajima F, Fukushi K (2011) Bioaugmented membrane bioreactor (MBR) with a GAC-packed zone for high rate textile wastewater treatment. Water Res 45 (6):2199–2206. https://doi.org/10.1016/j.watres.2011.01.013

Hai FI, Modin O, Yamamoto K, Fukushi K, Nakajiam F, Nghiem LD (2012) Pesticide removal by a mixed culture of bacteria and white-rot fungi. J Taiwan Inst Chem Eng 43(3):459–462. https://doi.org/10.1016/j.jtice.2011.11.002

Haq I, Kumar S, Raj A, Lohani M, Satyanarayana GNV (2017) Genotoxicity assessment of pulp and paper mill effluent before and after bacterial degradation using *Allium cepa* test. Chemosphere 169:642–650. https://doi.org/10.1016/j.chemosphere.2016.11.101

He Z, Xiao H, Tang L, Min H, Lu Z (2013) Biodegradation of di-n-butyl phthalate by a stable bacterial consortium, HD-1, enriched from activated sludge. Biores Technol 128:526–532. https://doi.org/10.1016/j.biortech.2012.10.107

Hu X, Li A, Fan J, Deng C, Zhang Q (2008) Biotreatment of p-nitrophenol and nitrobenzene in mixed wastewater through selective bioaugmentation. Biores Technol 99(10):4529–4533. https://doi.org/10.1016/j.biortech.2007.08.039

Hunger K, Gregory P, Miederer P, Berneth H, Heid C, Mennicke W (2004) Important chemical chromophores of dye classes. Indus Dyes Chem Properties Appl 13–112. doi:https://doi.org/10.1002/3527602011.ch2

Isik M, Sponza DT (2004) Monitoring of toxicity and intermediates of CI Direct Black 38 azo dye through decolorization in an anaerobic/aerobic sequential reactor system. J Hazard Mater 114 (1):29–39. https://doi.org/10.1016/j.jhazmat.2004.06.011

Jain N, Prajapati SK, Minocha AK, Verma CL (2001) Batch studies on the degradation of spent wash from distilleries. Indian J Environ Prot 21(2):122–126

Jain N, Minocha AK, Verma CL (2002) Degradation of predigested distillery effluent by isolated bacterial strains. doi:hdl.handle.net/123456789/17203

Jamil TS, Ghaly MY, El-Seesy IE, Souaya ER, Nasr RA (2011) A comparative study among different photochemical oxidation processes to enhance the biodegradability of paper mill wastewater. J Hazard Mater 185(1):353–358. https://doi.org/10.1016/j.jhazmat.2010.09.041

Kapdan IK, Tekol M, Sengul F (2003) Decolorization of simulated textile wastewater in an anaerobic–aerobic sequential treatment system. Process Biochem 38(7):1031–1037. https://doi.org/10.1016/S0032-9592(02)00238-8

Kasai Y, Kodama Y, Takahata Y, Hoaki T, Watanabe K (2007) Degradative capacities and bioaugmentation potential of an anaerobic benzene-degrading bacterium strain DN11. Environ Sci Technol 41(17):6222–6227. doi:https://doi.org/10.1021/es062842p

Kaushik G, Thakur IS (2009) Isolation of fungi and optimization of process parameters for decolorization of distillery mill effluent. World J Microbiol Biotechnol 25(6):955–964. https://doi.org/10.1007/s11274-009-9970-0

Kaushik G, Gopal M, Thakur IS (2010) Evaluation of performance and community dynamics of microorganisms during treatment of distillery spent wash in a three stage bioreactor. Biores Technol 101(12):4296–4305. https://doi.org/10.1016/j.biortech.2010.01.046

Knackmuss HJ (1996) Basic knowledge and perspectives of bioelimination of xenobiotic compounds. J Biotechnol 51(3):287–295. https://doi.org/10.1016/S0168-1656(96)01608-2

Kumar P, Chandra R (2006) Decolourisation and detoxification of synthetic molasses melanoidins by individual and mixed cultures of Bacillus spp. Biores Technol 97(16):2096–2102. https://doi.org/10.1016/j.biortech.2005.10.012

Kumar V, Wati L, Nigam P, Banat IM, Yadav BS, Singh D, Marchant R (1998) Decolorization and biodegradation of anaerobically digested sugarcane molasses spent wash effluent from biomethanation plants by white-rot fungi. Process Biochem 33(1):83–88. https://doi.org/10.1016/S0032-9592(97)00047-2

Kumar V, Dhall P, Kumar R, Singh YP, Kumar A (2012) Bioremediation of agro-based pulp mill effluent by microbial consortium comprising autochthonous bacteria. Sci World J 2012:1–7. https://doi.org/10.1100/2012/127014

Kumar V, Dhall P, Naithanl S, Kumar A, Kumar R (2014) Biological approach for the treatment of pulp and paper industry effluent in sequence batch reactor. J Bioremed Biodeg 5:218. https://doi.org/10.4172/2155-6199.1000218

Liang X, Devine CE, Nelson J, Sherwood Lollar B, Zinder S, Edwards EA (2013) Anaerobic conversion of chlorobenzene and benzene to CH4 and CO2 in bioaugmented microcosms. Environ Sci Technol 47(5):2378–2385. https://doi.org/10.1021/es3043092

Liebergott N, van Lierop B, Kovacs T, Nolin A (1990) A comparison of the order of addition of chlorine and chlorine dioxide in the chlorination stage. Tappi J 73(10):207–213

Lors C, Damidot D, Ponge JF, Périé F (2012) Comparison of a bioremediation process of PAHs in a PAH-contaminated soil at field and laboratory scales. Environ Pollut 165:11–17. https://doi.org/10.1016/j.envpol.2012.02.004

Lourenco ND, Novais JM, Pinheiro HM (2000) Reactive textile dye colour removal in a sequencing batch reactor. Water Sci Technol 42(5–6):321–328

Ma F, Guo JB, Zhao LJ, Chang CC, Cui D (2009) Application of bioaugmentation to improve the activated sludge system into the contact oxidation system treating petrochemical wastewater. Biores Technol 100(2):597–602. https://doi.org/10.1016/j.biortech.2008.06.066

Mackay D, Fraser A (2000) Bioaccumulation of persistent organic chemicals: mechanisms and models. Environ Pollut 110(3):375–391. https://doi.org/10.1016/S0269-7491(00)00162-7

Maes A, Van Raemdonck H, Smith K, Ossieur W, Lebbe L, Verstraete W (2006) Transport and activity of *Desulfito bacterium* dichloro eliminans strain DCA1 during bioaugmentation of 1, 2-DCA-contaminated groundwater. Environ Sci Technol 40(17):5544–5552. https://doi.org/10.1021/es060953i

Mahimairaja S (2004) Problems and prospects of agricultural use of distillery spentwash in India. Magnesium 1715:2100

Malaviya JP, Rathore VS (2007) Bioremediation of pulp and paper mill effluent by a novel fungal consortium isolated from polluted soil. Bioresour Technol 98(18):3647–3651. https://doi.org/10.1016/j.biortech.2006.11.021

Margesin R, Schinner F (1999) Biodegradation of diesel oil by cold-adapted microorganisms in presence of sodium dodecyl sulfate. Chemosphere 38(15):3463–3472. https://doi.org/10.1016/S0045-6535(98)00575-X

Mckague AB, Carlberg G (1996) Pulp bleaching and the environment. In: Dence CW, Reeve DW (eds) Pulp bleaching-principles and practice. Tappi Press, Atlanta, pp 746–820

McKenzie N, Yue S, Liu X, Ramsay BA, Ramsay JA (2014) Biodegradation of naphthenic acids in oils sands process waters in an immobilized soil/sediment bioreactor. Chemosphere 109:164–172. https://doi.org/10.1016/j.chemosphere.2014.02.001

Méndez-Paz D, Omil F, Lema JM (2005) Anaerobic treatment of azo dye acid orange 7 under fed-batch and continuous conditions. Water Res 39(5):771–778. https://doi.org/10.1016/j.watres.2004.11.022

Miranda MP, Benito GG, San Cristobal N, Nieto CH (1996) Color elimination from molasses wastewater by *Aspergillus niger*. Biores Technol 57(3):229–235. https://doi.org/10.1016/S0960-8524(96)00048-X

Misra SG, Mani D (1993) Ecosystem pollution. Indus Pub Co

Miyata N, Iwahori K, Fujita M (1998) Manganese-independent and-dependent decolorization of melanoidin by extracellular hydrogen peroxide and peroxidases from Coriolus hirsutus pellets. J Ferment Bioeng 85(5):550–553. https://doi.org/10.1016/S0922-338X(98)80107-4

Mohan D, Singh KP, Sinha S, Gosh D (2004) Removal of pyridine from aqueous solution using low cost activated carbons derived from agricultural waste materials. Carbon 42(12):2409–2421. https://doi.org/10.1016/j.carbon.2004.04.026

Mohan SV, Rao NC, Prasad KK, Sarma PN (2005) Bioaugmentation of an anaerobic sequencing batch biofilm reactor (AnSBBR) with immobilized sulphate reducing bacteria (SRB) for the treatment of sulphate bearing chemical wastewater. Process Biochem 40(8):2849–2857. https://doi.org/10.1016/j.procbio.2004.12.027

Mojiri A, Aziz1 HA, Aziz SQ (2013) Trends in physical-chemical methods for landfill leachate treatment. Inter J Sci Res Environ Sci 1(2):16–25

Moutaouakkil A, Zeroual Y, Dzayri FZ, Talbi M, Lee K, Blaghen M (2004) Decolorization of azo dyes with *Enterobacter agglomerans* immobilized in different supports by using fluidized bed bioreactor. Curr Microbiol 48(2):124–129. https://doi.org/10.1007/s00284-003-4143-0

Ngah WW, Teong LC, Hanafiah MAKM (2011) Adsorption of dyes and heavy metal ions by chitosan composites: a review. Carbohyd Polym 83(4):1446–1456. https://doi.org/10.1016/j.carbpol.2010.11.004

Nyhoim N (1996) Biodegradability characterization of mixtures of chemical contaminants in waste water-the utility of biotests. Water Sci Technol 33(6):195–206. https://doi.org/10.1016/0273-1223(96)00326-5

O'Neill C, Lopez A, Esteves S, Hawkes FR, Hawkes DL, Wilcox S (2000) Azo-dye degradation in an anaerobic-aerobic treatment system operating on simulated textile effluent. Appl Microbiol Biotechnol 53(2):249–254. https://doi.org/10.1007/s002530050016

Padoley KV, Mudliar SN, Banerjee SK, Deshmukh SC, Pandey RA (2011) Fenton oxidation: a pretreatment option for improved biological treatment of pyridine and 3-cyanopyridine plant wastewater. Chem Eng J 166(1):1–9. https://doi.org/10.1016/j.cej.2010.06.041

Paliwal R, Uniyal S, Rai JPN (2015) Evaluating the potential of immobilized bacterial consortium for black liquor biodegradation. Environ Sci Pollut Res Int 22(9):6842–6853. https://doi.org/10.1007/s11356-014-3872-x

Patel H, Madamwar D (2002) Effects of temperatures and organic loading rates on biomethanation of acidic petrochemical wastewater using an anaerobic up flow fixed-film reactor. Biores Technol 82(1):65–71. https://doi.org/10.1016/S0960-8524(01)00142-0

Pendashteh AR, Fakhru'l-Razi A, Chuah TG, Radiah ABD, Madaeni SS, Zurina ZA (2010) Biological treatment of produced water in a sequencing batch reactor by a consortium of isolated halophilic microorganisms. Environ Technol 31(11):1229–1239. doi:https://doi.org/10.1080/09593331003646612

Pokhrel D, Viraraghavan T (2004) Treatment of pulp and paper mill wastewater—a review. Sci Total Environ 333(1–3):37–58. https://doi.org/10.1016/j.scitotenv.2004.05.017

Quan X, Shi H, Liu H, Wang J, Qian Y (2004) Removal of 2,4-dichlorophenol in a conventional activated sludge system through bioaugmentation. Process Biochem 39(11):1701–1707. https://doi.org/10.1016/S0032-9592(03)00307-8

Quero GM, Cassin D, Botter M, Perini L, Luna GM (2015) Patterns of benthic bacterial diversity in coastal areas contaminated by heavy metals, polycyclic aromatic hydrocarbons (PAHs) and polychlorinated biphenyls (PCBs). Front Microbiol 6. doi:https://doi.org/10.3389/fmicb.2015.01053

Ravanchi MT, Kaghazchi T, Kargari A (2009) Application of membrane separation processes in petrochemical industry: a review. Desalination 235(1):199–244. https://doi.org/10.1016/j.desal.2007.10.042

Richardson RE (2013) Genomic insights into organohalide respiration. Curr Opin Biotechnol 24 (3):498–505. https://doi.org/10.1016/j.copbio.2013.02.014

Richardson SD, Plewa MJ, Wagner ED, Schoeny R, DeMarini DM (2007) Occurrence, genotoxicity and carcinogenicity of regulated and emerging disinfection by-products in drinking water: a review and roadmap for research. Mutat Res 636:178–242. https://doi.org/10.1016/j.mrrev.2007.09.001

Robinson T, McMullan G, Marchant R, Nigam P (2001) Remediation of dyes in textile effluent: a critical review on current treatment technologies with a proposed alternative. Biores Technol 77(3):247–255. https://doi.org/10.1016/S0960-8524(00)00080-8

Sahoo DK, Gupta R (2005) Evaluation of ligninolytic microorganism for efficient decolourization of a small pulp and paper mill effluent. Proc Biochem 40(5):1573–1578. https://doi.org/10.1016/j.procbio.2004.05.013

Saranya P, Ramani K, Sekaran G (2014) Biocatalytic approach on the treatment of edible oil refinery wastewater. RSC Adv 4(21):10680–10692. https://doi.org/10.1039/C3RA43668C

Satyawali Y, Balakrishnan M (2008) Wastewater treatment in molasses-based alcohol distilleries for COD and color removal: a review. J Environ Manage 86(3):481–497. https://doi.org/10.1016/j.jenvman.2006.12.024

Savant DV, Abdul-Rahman R, Ranade DR (2006) Anaerobic degradation of adsorbable organic halides (AOX) from pulp and paper industry wastewater. J Bior Technol 97:1092–1104. https://doi.org/10.1016/j.biortech.2004.12.013

Schoefs O, Perrier M, Samson R (2004) Estimation of contaminant depletion in unsaturated soils using a reduced-order biodegradation model and carbon dioxide measurement. Appl Microbiol Biotechnol 64(1):53–61. https://doi.org/10.1007/s00253-003-1423-3

Sekiguchi Y (2006) Yet-to-be cultured microorganisms relevant to methane fermentation processes. Microbes Environ 21(1):1–15. https://doi.org/10.1264/jsme2.21.1

Sekiguchi Y, Kamagata Y (2004) Microbial community structure and functions in methane fermentation technology for wastewater treatment. Strict Facultative Anaerobes Med Environ Aspects 361–384

Sekiguchi Y, Kamagata Y, Harada H (2001) Recent advances in methane fermentation technology. Curr Opin Biotechnol 12(3):277–282. https://doi.org/10.1016/S0958-1669(00)00210-X

Sharma A, Thakur IS (2009) Identification and characterization of integron mediated antibiotic resistance in pentachlorophenol degrading bacterium isolated from the chemostat. J Environ Sci 21(6):858–864. https://doi.org/10.1016/S1001-0742(08)62353-0

Sharma P, Singh L, Mehta J (2010) COD reduction and colour removal of simulated textile mill wastewater by mixed bacterial consortium. Rasayan J Chem 3(4):731–735

Shaw LE, Lee D (2009) Sonication of pulp and paper effluent. Ultrasonicssonochemistry 16 (3):321–324. https://doi.org/10.1016/j.ultsonch.2008.10.011

Singh P (2007) Sequential anaerobic and aerobic treatment of pulp and paper mill effluent in pilot scale bioreactor. J Environ Biol 28(1):77–82

Singh P, Thakur IS (2006) Colour removal of anaerobically treated pulp and paper mill effluent by microorganisms in two steps bioreactor. Bioresour Technol 97(2):218–223. https://doi.org/10.1016/j.biortech.2005.02.022

Singh S, Chandra R, Patel DK, Reddy MMK, Rai V (2008) Investigation of the biotransformation of pentachlorophenol and pulp paper mill effluent decolorisation by the bacterial strains in a mixed culture. Bioresour Technol 99(13):5703–5709. https://doi.org/10.1016/j.biortech.2007.10.022

Singhal A, Thakur IS (2009) Decolourization and detoxification of pulp and paper mill effluent by *Cryptococcus* sp. Biochem Eng J 46(1):21–27. https://doi.org/10.1016/j.bej.2009.04.007

Soloman PA, Basha CA, Velan M, Balasubramanian N, Marimuthu P (2009) Augmentation of biodegradability of pulp and paper industry wastewater by electrochemical pre-treatment and optimization by RSM. Sep Purif Technol 69(1):109–117. https://doi.org/10.1016/j.seppur.2009.07.002

Sorensen SR, Holtze MS, Simonsen A, Aamand J (2007) Degradation and mineralization of nanomolar concentrations of the herbicide dichlobenil and its persistent metabolite 2,6-dichlorobenzamide by *Aminobacter* spp. isolated from dichlobenil-treated soils. Appl Environ Microbiol 73(2):399–406. https://doi.org/10.1128/AEM.01498-06

Srivastava S, Thakur IS (2007) Evaluation of biosorption potency of *Acinetobacter* sp. for removal of hexavalent chromium from tannery effluent. Biodegradation 18(5):637–646. https://doi.org/10.1007/s10532-006-9096-0

The Energy and Resources Institute (TERI) Policy report (2015) New Delhi—110 003, INDIA

Tiku DK, Kumar A, Chaturvedi R, Makhijani SD, Manoharan A, Kumar R (2010) Holistic bioremediation of pulp mill effluents using autochthonous bacteria. Int Biodeter Biodegrad 64 (3):173–183. https://doi.org/10.1016/j.ibiod.2010.01.001

Tran NH, Urase T, Ngo HH, Hu J, Ong SL (2013) Insight into metabolic and cometabolic activities of autotrophic and heterotrophic microorganisms in the biodegradation of emerging trace organic contaminants. Bioresour Technol 146:721–731. https://doi.org/10.1016/j.biortech.2013.07.083

Trejo-Hernandez MR, Lopez-Munguia A, Ramirez RQ (2001) Residual compost of *Agaricus bisporus* as a source of crude laccase for enzymic oxidation of phenolic compounds. Process Biochem 36(7):635–639. doi:https://doi.org/10.1016/S0032-9592(00)00257-0

Tyagi M, da Fonseca MMR, de Carvalho CC (2011) Bioaugmentation and biostimulation strategies to improve the effectiveness of bioremediation processes. Biodegradation 22(2):231–241. https://doi.org/10.1007/s10532-010-9394-4

Van der Zee FP, Villaverde S (2005) Combined anaerobic–aerobic treatment of azo dyes—a short review of bioreactor studies. Water Res 39(8):1425–1440. https://doi.org/10.1016/j.watres.2005.03.007

Veeresh GS, Kumar P, Mehrotra I (2005) Treatment of phenol and cresols in upflow anaerobic sludge blanket (UASB) process: a review. Water Res 39(1):154–170. https://doi.org/10.1016/j.watres.2004.07.028

Wesenberg D, Kyriakides I, Agathos SN (2003) White-rot fungi and their enzymes for the treatment of industrial dye effluents. Biotechnol Adv 22(1):161–187. doi:https://doi.org/10.1016/j.biotechadv.2003.08.011

White PA, Rasmussen JB, Blaise C (1996) Comparing the presence, potency, and potential hazard of genotoxins extracted from a broad range of industrial effluents. Environ Mol Mutagen 27 (2):116–139. https://doi.org/10.1002/(SICI)

Zhan XM, Rodgers M, O'Reilly E (2006) Biofilm growth and characteristics in an alternating pumped sequencing batch biofilm reactor (APSBBR). Water Res 40(4):817–825. https://doi.org/10.1016/j.watres.2005.12.003

Zhang JL, Qiao CL (2002) Novel approaches for remediation of pesticide pollutants. Int J Environ Pollut 18(5):423–433. https://doi.org/10.1504/IJEP.2002.002337

Zhao LJ, Ma F, Guo JB (2009) Applicability of anoxic-oxic process in treating petrochemical wastewater. J Zhejiang Univ Sci A 10(1):133–141. https://doi.org/10.1631/jzus.A0820006

Chapter 12
Fluoride Remediation from Drinking Water

Neelam Rawat and Vinay Kumar Patel

Abstract Fluoride in potable water imposes intense effects to teeth and bones. In this chapter, the adverse effects of fluoride contaminants in drinking water and their remediation has been discussed. The fluoride remediation has been largely covered in two sections reflecting the membrane techniques (reverse osmosis, nanofiltration, dialysis and electrodialysis) and adsorption techniques (using adsorbents like alumina/aluminium-based materials, clays and soils, calcium-based minerals, synthetic compounds and carbon-based materials). In addition, some new remarkable advancement in the fluoride remedial methods made in current year has been discussed.

Keywords Fluoride · Adsorption · Reverse osmosis · Membrane
Drinking water

12.1 Introduction

Water is the most vital element for sustaining life on the earth. It is a natural resource which is in abundance, i.e. about 71% of earth's surface, among which oceans itself holds 96.5%. Fresh water holds only 2.5% among which only 1% (surface and ground water) is accessible to us as drinking water. But in today's era of urbanization and industrialization, water resources are getting polluted, and hence, the availability as well as the quality of drinking water in developing countries is degrading day by day.

Usually drinking water collected from surface water resources (wells, rivers, ponds, lakes, etc.) along with groundwater resources is biocontaminated and also have the presence of toxic anions (fluoride, arsenic, arsenite, nitrate, sulphate, phosphate, chromate and selenate) which causes many diseases.

N. Rawat · V. K. Patel (✉)
Department of Mechanical Engineering, Govind Ballabh Pant Institute of Engineering and Technology, Pauri Garhwal 246001, Uttarakhand, India
e-mail: vinaykrpatel@gmail.com

S. Bhattacharya et al. (eds.), *Water Remediation*, Energy, Environment, and Sustainability, https://doi.org/10.1007/978-981-10-7551-3_12

Fluorine is a highly reactive electronegative element which has an extraordinary affinity with a positively charged anion like calcium and magnesium. It is found in the form of sellate (MgF_2), cryolite (Na_3AlF_6), fluorspar (CaF_2) and fluorapatite [$3Ca_3(PO_4)_2Ca(F, Cl_2)$]. Fluoride enters into the groundwater by various mediums like natural weathering, precipitation and leaching, besides anaesthetics, antibiotics, fertilizers, pesticides (containing NaF), contaminated beverages/food products and tooth paste industry are the major source of fluoride pollution in water.

Fluoride in drinking water is beneficial if its concentration is between 1 and 1.5 mg/L as specified by WHO (1984) and environmental protection agency (EPA). It helps in calcification of dental enamel and maintains healthy bones. But if the consumption of fluoride increases above the permissible limits (>1.5 mg/L), it results in dental fluorosis and/or skeletal fluorosis (Sorg 1978). Besides fluorosis, excessive ingestion of fluoride also causes muscle fibre degeneration, low haemoglobin levels, thyroid, gastrointestinal problems, abdominal pain, reduced immunity, etc. In addition to the health problems to humans, the continuous use of high-fluoride concentrate water in agricultural fields may adversely affect the crop growth. Hence, it is imperative to lower down the fluoride level to permissible limits from our water resources.

12.2 History and Remediation Processes

With the rapid growth in industrialization, water resources flooded with high level of fluoride pose a great concern to sustainable growth of plants and animals. It has been found that countries like USA, Africa, Thailand, Sri Lanka, India, Arabia, Iran, Iraq, Kenya, Syria, etc., has high level of fluoride in groundwater, i.e. up to 30 mg/L or more (Mameri et al. 1998). Primarily there are three main solutions that can minimize the problem of excessive fluoride in water supply and they are:

(a) By the use of alternative water resources.
(b) Dilution, i.e. mixing the contaminated water to the clean water to attain the requisites.
(c) Defluoridation, i.e. removing excess of fluoride from drinking water.

The first two solutions, i.e. alternative water resources (like surface water, rainwater and low fluoride ground water) and dilution alone cannot cater the high demand of portable drinking water due to its insufficiency. So defluoridation of drinking water is the only practical remedial solution of the problem of high-fluoride contents in potable water. Fundamentally, the defluoridation techniques can roughly be divided into:

1. Precipitation and coagulation
2. Adsorption techniques and
3. Membrane/layer techniques.

12.2.1 Precipitation and Coagulation

This technique (Fig. 12.1) comprises the inclusion of coagulants (such as alum and lime) into the water which agglomerate the fluoride particles thereby precipitating it. This process occurs in two stages, i.e. firstly on addition of lime, precipitate is formed followed by addition of alum in second stage causing coagulation.

Bulusu et al. (1979) of National Environmental Engineering Research Institute, Nagpur, India, demonstrated a technique popularized as Nalgonda technique, in which water is treated with alkali, chlorine and aluminium sulphate or aluminium chloride or both to remove fluoride from drinking water. This process is cost effective, and so, it is used at large scale in India. The Nalgonda process involves the addition of lime and alum followed by flocculation or coagulation, precipitation and filtration processes. Then, this treated water is passed through a microfiltration membrane which physically separates metal contaminant particles from the wastewater and the solids from the membrane are removed by periodically back flushing it.

Advantages

- It is an established method.
- It is widely used in community level due to its low cost and effectiveness.

Limitations

- Inability to use in high-fluoride-level contamination.
- Large quantity of aluminium sulphate is needed (up to 0.7–1.2 g/L)
- High level of aluminium in treated water causes adverse health effects, since aluminium has neurotoxic property.
- Skilled manpower is needed.

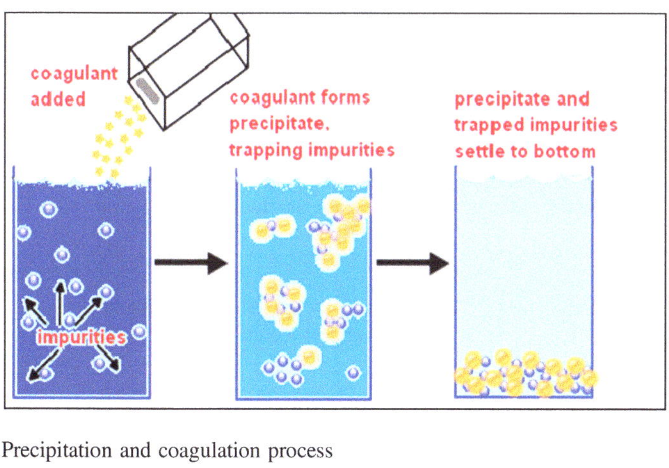

Fig. 12.1 Precipitation and coagulation process

12.2.2 Adsorption Techniques

Adsorption is a process of addition of atoms or ions to the adsorbent's surface from a gas, liquid or dissolved solid. Since past few years, various adsorbent materials are being utilized for defluoridation of potable water using economical and effective adsorbing agent. Surface adsorption process has been an interesting area of research for defluoridation of water for a few decades and has been frequently used in water treatment plants in a large scale. Adsorption technique is widely used due to its low cost, easy of operate, high reusability prospects and easy availability of adsorbent. Most commonly used adsorbents are activated alumina, activated carbon, activated alumina-coated silica gel, calcite, activated saw dust, activated coconut shell carbon and activated fly ash, groundnut shell, coffee husk, rice husk, agnesia, serpentine, tricalcium phosphate, bone charcoal, activated soil sorbent, carbon, defluoron-1, efluoron-2, etc. (Kariyanna 1987; Barbier and Mazounie 1984; Muthukumaran et al. 1995; Rongshu et al. 1995; Min et al. 1999; Wang and Reardon 2001; Nava et al. 2003; Padmavathy et al. 2003; Thergaonkar and Nawalakhe 1971).

Fan et al. (2007) have described the process of adsorption on the solid particles takes place in three steps and they are:

1. External mass transfer: fluoride (F^-) ions diffuse/transport to the outer surface of the adsorbent from bulk solution across the boundary layer surrounding the adsorbent particle;
2. Fluoride ions adsorb on to the surfaces of particle;
3. It is probable that the adsorbed fluoride ions exchange with the structural elements inside adsorbent particles as per the chemistry of solids, or there is transfer of adsorbed fluoride ions to the internal surfaces for porous materials (intra-particle diffusion).

Table 12.1 indicates the availability and their efficiency/economy of some of the commonly available absorbents for fluoride removal. Boruff (1934) was among the first researchers who have worked in the area of removing excess fluoride level from potable water (Davey 1939; Fink and Lindsay 1936; Adler et al. 1938). He tested aluminium sulphate, sodium aluminate, zeolite, activated alumina and bauxite as aluminium-based materials and also explored the appropriateness of common adsorbents comprising silica gel, sodium silicate and ferric salts.

Table 12.1 Accessibility of common adsorbents for fluoride elimination (Ndé-Tchoupé et al. 2015)

Material	Comments	Availability
Metal oxides and hydroxides	Reveal relatively low efficiency without physico-chemical modification	Low
Biosorbents		Very high
Geomaterials		High
Carbonaceous materials, industrial by-products	Typically need costly physico-chemical activation	Low
Bone char	Produced by carbonizing animal bones	High

12.2.2.1 Activated Alumina and Aluminium Compounds

Activated alumina is most widely used adsorbent for removing fluoride from water. Savinelli and Black have studied the activated alumina for removal of fluoride a long back (Sovinelli and Black 1958). Johnston and Heijnen (2002) demonstrated the selectivity sequence of anion adsorption on activated alumina in the pH range of 5.5–8.5 as following:

$$OH^- > H_2AsO_4^- > Si(OH)_3O^- > HSeO_3^- > F^- > SO_4^- > CrO_4^- > HCO_3^- > Cl^- > NO_3^- > Br^- > I^-$$

The only disadvantage of activated alumina as an adsorbent is that it works only on pH below 6 which is not suitable for practical applications. Also, it has been reported that alumina causes adverse effect on human health by causing Alzheimer's disease and has other health hazards too (Davison et al. 1982; Crapper et al. 1973; Miller et al. 1984; Martyn et al. 1989). The efficiency of activate alumina for removing fluoride is affected by the hardness, pH and surface loading of alumina adsorbent.

In recent years, researches have been done to use impregnated alumina with low-cost porous solids. Lanthanum (III)- and ytterbium (III)-impregnated alumina has shown satisfactory results for the removal of fluoride from the drinking water (Wasay et al. 1996). Tripathy et al. (2006) has investigated the fluoride removal from water through adsorption by using alum-impregnated activated alumina (AIAA) and found that the removal of fluoride increases with increase in pH up to 6.5 and also that AIAA can remove fluoride up to 0.2 mg/L from water having 20 mg/L of fluoride, i.e. about 92.6%.

12.2.2.2 Activated Carbon

The most commonly used adsorbents are activated alumina and activated carbon. Mckee and Johnston examined the utilization of powdered activated carbon formulated from various raw materials for fluoride removal and found satisfactory performance (Mckee and Johnston 1934). The process was found to be dependent on pH displaying good results only at pH 3 or less. Therefore, the pH adjustment makes this material and process expensive. Many researchers has worked on evolving new adsorbents which uses metal ions for defluoridation, and it has been concluded that aluminium impregnated carbon is 3–5 times more effective in eliminating fluoride from drinking water as compared to plain activated carbon (Ramos et al. 1999). When manganese oxide (MnO)-coated granular activated carbon adsorbent was applied for the removal of fluoride from water, its adsorption capacity was found to be enhanced by three times as compared to that of uncoated granular activated carbon (Ma et al. 2009).

12.2.2.3 Zeolites

Zeolite occurs naturally in crystalline alumina-silicates, having a tetrahedron structure which has four oxygen atoms attached to relatively small silicon or aluminium atom. They can be produced synthetically as well. Zeolite has an ion exchange property which makes it suitable for waste water treatment for removal of contaminants. Clinoptilolite is the most commonly known and used zeolite in pollute control (Akgul et al. 2006; Karadag et al. 2008). In order to enhance the adsorption and catalysis properties of zeolites, modifications are performed on it. These modifications are done by using three techniques, viz. skeleton element, non-skeletal element and surface modification, among which non-skeleton element modification method is the most widely used method to modify the zeolite. The basic zeolite process is shown in Fig. 12.2. This is done by using ion exchange or impregnation method by treating the zeolite in the inorganic acids or salts of metal (Bowman 2003; Pang et al. 2007; Hu et al. 1999). Natural STI zeolite was modified by Fe (III) ion and batch sorption studies were conducted for defluoridation of water having maximum adsorption capacity of 2.31 mg/g. The studies showed that fluoride level can be reduced to 1 mg/L under optimum conditions. Exhausted Fe (III)-STI zeolite was then regenerated by using 1 mol/L HCl eluent, and it was found that regenerated samples has remained with good adsorptive performances (Sun et al. 2011).

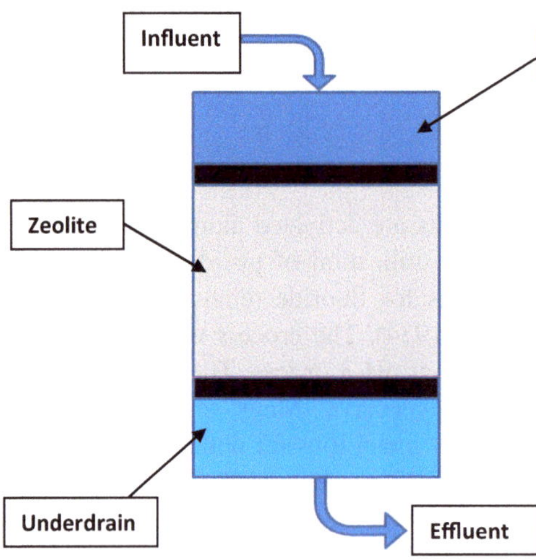

Fig. 12.2 Basic zeolite process

12.2.2.4 Geomaterials

Das et al. (2005) investigated the adsorption of fluoride on thermally activated titanium-rich bauxite (TRB). The adsorption capacity of TRB was found to be greatly enhanced when it was thermally activated at moderate temperatures in the range of 300–450 °C. Adsorption process was examined to be rapid with maximum level accomplished within 90 min. The uptake of fluoride was reported to increase with increasing pH, being a maximum at pH range of 5.5–6.5, and thereafter, the uptake was found decreasing. The uptake process of fluoride from drinking water was found independent of the common interfering ions present in the water, thus clearly indicating the fluoride-specific sorption behaviour of TRB. Nickel laterites and chromite mine overburden commonly comprise high content of iron and small amounts of alumina, chromium, cobalt, nickel, manganese. Due to their high iron content in the form of goethite, Sarkar et al. studied the mechanistic function of laterite in removal of fluoride and demonstrated that the mechanism of fluoride adsorption is dictated by the zero point charge of laterite, and the pH plays a very vital role through influencing the surface characteristics of laterite (Sarkar et al. 2006).

12.2.2.5 Bone Char

Bone char is found to be the most affordable and efficient substitute to natural tourmaline and activated alumina for fluoride removal and the experiments shows that the fluoride removal rate is greater in it. Bone char column adsorption experiments also specify that the fluoride elimination tendency is affected by the flow rate and bed height. Although it has many benefits, like in addition to fluoride it can remove other soluble contaminants like arsenate, dyes and heavy metals (Rojas-Mayorga et al. 2015) and low cost, it still is not widely used for removing fluoride from drinking water because of its unpleasant taste and fresh bone handling problems (Jagtap et al. 2012).

12.2.2.6 Low-Cost Materials

Other natural low-cost materials that can remove fluoride from drinking water are kaolinite, lignite, charfines, bentonite and nirmali seeds, etc. Researches have been done to get the best alternative which has the highest fluoride removal rate and that can also be used in practical applications. It was found that bentonite and charfines have approximately same fluoride removal rates, but charfines is considered for fluoride removal since it can work in both batch as well as continuous processes as compared to bentonite which can only be used in continuous process Srimurali et al. (1998). Srimurali et al. have also demonstrated that the fluoride elimination increases with time and then approaches a constant value, also with the size of adsorbent it increases as the size decreases. The fluoride elimination rate is dependent on the pH of the water, and as the pH increases the efficiency of fluoride removal decreases while it has no effect on the chemical pretreatment of sorbent Srimurali et al. (1998).

12.2.2.7 Miscellaneous Materials

Mohan et al. have studied the use of a green waste-based biochar for the removal of fluoride from the drinking water which was obtained as a by-product of the biooil production. In the experiment, they used pine wood as well as pine bark which was first pyrolyzed in auger-fed reactor (1 kg/h) at a temperature of 400 and 450 °C and then added to water having a pH range of 2–10. The adjustment of pH of water was made by using H_2SO_4. The batch sorption studies of biochars for the elimination of fluoride showed best results at pH 2 while it goes on decreasing on higher pH levels. The fluoride elimination from pine wood char was found higher than that of pine bark char. Fluoride sorption is well described by Langmuir adsorption model (2012). Also it has been concluded that the fluoride sorption capabilities of biochars are greater than or comparable to that of activated carbons and many other materials used till date for the defluoridation of water. The use of corn stover biochar is also seen in the research of Mohan et al. in which he prepared the biochar by slow pyrolysis at 500 °C. It was then magnetized by mixing aqueous biochar suspension with aqueous ferrous/ferric solution, followed by aqueous NaOH treatment. Both corn stover biochar (CSBC) and magnetic corn stover biochar (MCSBC) indicated the successful treatment of fluoride contaminated water. Also a decrease in adsorption capacity was found in both CSBC and MCSBC with the increase in temperature (2014).

Mohapatra et al. demonstrated the synthesis and application of additive-assisted nanogoethite (α-FeOOH). Goethite is an iron bearing oxide mineral which is widely used as an adsorbent because of adsorption of oxyanions and cations in its complex matrix. In the study, Mohapatra used hydrazine sulphate additive to assist the nanosized goethite particles. In the experiment, it was found that on increasing the Cl^- and sulphate ion concentration, the adsorption of fluoride decreased. Also the fluoride concentration from the water can be reduced to 0.5 mg/L when batch sorption process was used in three stages having a pH range of 6–8. The maximum adsorption capacity of about 59 mg/L of goethite was observed (2010).

Bleaching powder being a disinfectant has also been applied as an adsorbent materials for fluoride removal from water. At pH 6–10, defluoridation from water occurred due to adsorbent behaviour of bleaching powder as well as by precipitation in the form of calcium fluoride, at a dose of 50 g/L (Kagne et al. 2009). Besides all these adsorbents, some researchers have also examined the different types of adsorbents like potassium permanganate modified carbon derived from steam pyrolysis of rice straw, hybrid thorium phosphate composite, granular acid treated bentonite, etc., for the elimination of fluoride from the potable water (Daifullah et al. 2007; Islam et al. 2011; Ma et al. 2011).

Despite of using adsorption technique over past 80 years, and having advantages like low-cost treatment and effective fluoride removal (i.e., up to 1.5 mg/L), there are still some limitations of this technique which has been listed by (Maheshwari 2006) and these are as follows.

Limitations

- The procedure is greatly pH dependent, i.e., operates well only to pH of 5–6.
- High concentration of salts and TDS in water can lead to fouling of the alumina bed.
- Interference of other ions like sulphate, phosphate, carbonate, etc., in adsorption of fluoride ion.
- This technique has poor integrity, low adsorption capacity and also there is a need of pretreatment.
- Constant needs of regeneration in every 4–5 months and also after each regeneration the effectiveness of absorbent decreases.

12.2.3 Membrane Techniques

Membrane techniques are those which use a semipermeable membrane to separate the ions from the water. Fundamentally, it is divided into three main sections, viz. reverse osmosis (RO), nanofiltration (NF) and dialysis and electrodialysis processes.

12.2.3.1 Reverse Osmosis

Although there are many conventional techniques to purify the water which are being used since long ago and considered to be cost effective, due to their inefficiency to work on higher pH level (>6), non-user friendly nature and long payback period new techniques are being used nowadays. Reverse osmosis (RO) membrane technique gives water of extreme water purity level (Arora et al. 2004; Fu et al. 1995; Schneiter and Middlebrooks 1983). RO is the technique in which the soluble contaminants are eliminated from water by applying a pressure which overcomes the natural osmotic pressure, and hence, forces the clean water to pass through the semipermeable membrane by restricting the flow of contaminants. Ndiaye et al. (2005) demonstrated fluoride removal with RO technique and found that fluoride removed was higher than 98% with consideration that RO membranes were fully regenerated after each set of experiments. The determining factors for the selection of membrane of RO are cost, recovery, rejection, raw water characteristics and pretreatment. The efficiency of the RO system is greatly influenced by the pressure, temperature, raw water characteristics and regular monitoring and maintenance, etc (Fig. 12.3).

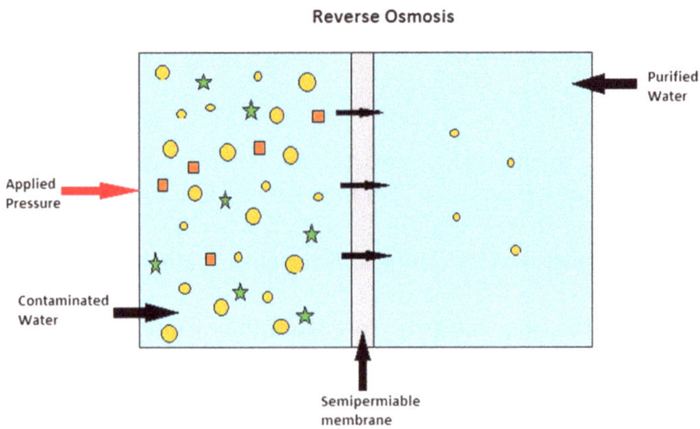

Fig. 12.3 Reverse osmosis process

12.2.3.2 Nanofilteration (NF)

NF is similar in operation to RO technique with the slight difference that in NF membranes there are comparatively larger pores which in turn lowers the applied pressure to remove the fluoride content from the water. NF has faster flow rate and is less exhaustive process than RO.

In past years, the membrane technology was not so popular due to its high costs but now, with the advancement in technology and increasing demands of safe drinking water it is being widely used. Furthermore, the operational and capital costs of the RO plant are now decreasing with the increase in plant capacity (Babra et al. 1997). Many researchers have (Arora et al. 2004; Drioli et al. 1999) studied the use of membrane technique for portable water production and have come up with many conclusions. Some of the advantages and limitation are as follows:

Advantages

- The membrane technique is highly effective with an another advantage that the membrane can trap other contaminants like pesticides, microbes, heavy metals, etc., from the water.
- It is a continuous process for water supply.
- Since it is a physical process, so there is no need of chemicals.
- Membranes generally have a long life so regeneration or replacement is not needed frequently.
- It also works under a wide-ranging pH.
- The interference by other ions has not been found.
- The process is simple, reliable and fully automated with minimum or no manpower requirement.

Limitations

- It eliminates all the ions including essential minerals present in water, so remineralization is also required after treatment.
- The process is relatively expensive.
- Most of the water gets wasted as brine, and hence, not suitable for water having high salts and TDS (total dissolved solids).
- Also the disposal of brine is a problem.

12.2.3.3 Dialysis and Electrodialysis

Dialysis is a process in which the separation of solutes present in a liquid is driven by transport of solute through the membrane instead of retaining the solutes (Donnan effect) while water passes through it as in RO and NF (Donnan 1995). In electrodialysis the ionic components from the water are removed through ion exchange membranes under the influence of an electric field. Hichour (Hichour et al. 2000) investigated the Donnan dialysis process in a counter current flow arrangement in which the anion-exchange membrane was loaded with NaCl and the feed was NaF together with other sodium salts, and it was found that fluoride migrated to the receiver while other ions to the feed.

12.3 Conclusions

Presence of fluoride in the drinking water is a major concern to the health of the mankind. So, it is of utmost importance that a research group should identify the source and cause of contaminants in the drinking water, and henceforth, find the most effective remedy process to solve this problem. Also the effective methods are to be developed which will help both urban as well as rural areas. In view of selection of appropriate method considering its suitability, efficacy, economy, and an overview of different processes have been discussed.

References

Adler H, Klein G, Lindsay FK (1938) Removal of fluorides from potable water by tricalcium phosphate. Ind Eng Chem 30:163–165

Akgul M, Karabakan A, Acar O, Yurum YD (2006) Removal of silver (I) from aqueous solutions with clinoptilolite. Micropor Mesopor Mat 94:99–104

Arora M, Maheshwari RC, Jain SK, Gupta A (2004) Use of membrane technology for potable water production. Desalination 170:105–112

Babra D, Caputi P, Cifoni DS (1997) Drinking water supply in Italy. Desalination 113:111–117

Barbier JP, Mazounie P (1984) Methods of reducing high fluoride content in drinking water, Fluoride removal methods—filtration through activated alumina: a recommended technique. Water Supply 2:3–4

Boruff CS (1934) Removal of fluorides from drinking waters. Ind Eng Chem 26:69–71

Bowman RS (2003) Applications of surfactant-modified zeolites to environmental remediation. Micropor Mesopor Mat 61:43–56

Bulusu KR, Sundaresan BB, Pathak BN, Nawlakhe WG, Kulkarni DN, Thergaonkar VP (1979) Fluorides in water, defluoridation methods and their limitations. J Inst Eng (India)-Environ Eng Div 60:1–25

Crapper DR, Krishnan SS, Dalton AJ (1973) Brain aluminum distribution in Alzheimer's disease and experimental neurofibrillary degeneration. Science 180:511–513

Daifullah AAM, Yakout SM, Elreefy SA (2007) Adsorption of fluoride in aqueous solutions using $KMnO_4$-modified activated carbon derived from steam pyrolysis of rice straw. J Hazard Mater 147:633–643

Das N, Pattanaik P, Das R (2005) Defluoridation of drinking water using activated titanium rich bauxite. J Colloid Interface Sci 292:1–10

Davey WB (1939) The use of bone and other phosphates for the removal of fluorine from drinking water. Master's Thesis, University of Arizona, Tucson, AZ, USA

Davison AM, Oli H, Walker GS, Lewins AM (1982) Water supply aluminium concentration, dialysis dementia, and effect of reverse-osmosis water treatment. The Lancet 320:785–787

Donnan FG (1995) Theory of membrane equilibria and membrane potentials in the presence of non-dialysing electrolytes. A contribution to physical-chemical physiology. J Membr Sci 100:45–55

Drioli E, Lagana F, Crlscuoh A, Barbieri G (1999) Integrated membrane operations in desalination processes. Desalination 122:141–145

Fan J, Xu Z, Zheng S (2007) Comment on Factors influencing the removal of fluoride from aqueous solution by calcined Mg–Al–CO3 layered double hydroxides. J Hazard Mater 139:175–177

Fink GJ, Lindsay FK (1936) Activated alumina for removing fluorides from drinking water. Ind Eng Chem 28:947–948

Fu P, Ruiz H, Lozier J, Thompson K, Spangenberg C (1995) A pilot study on groundwater natural organics removal by low-pressure membranes. Desalination 102:47–56

Hichour M, Persin F, Sandeaux J, Gavach C (2000) Fluoride removal from waters by Donnan dialysis. Sep Purif Technol 18:1–11

Hu ZH, Wei LH, Dong JX, Wang YR, Chen SY, Peng SY (1999) Modification of the external surface of ZSM-5 by a metal surfactant. Micropor Mesopor Mat 28:49–55

Islam M, Mishra PC, Patel R (2011) Fluoride adsorption from aqueous solution by a hybrid thorium phosphate composite. Chem Eng J 166:978–985

Jagtap S, Yenkie MK, Labhsetwar N, Rayalu S (2012) Fluoride in drinking water and defluoridation of water. Chem Rev 112:2454–2466

Johnston R, Heijnen H (2002) Safe water technology for arsenic removal. Report, World Health Organization (WHO)

Kagne S, Jagtap S, Thakare D, Devotta S, Rayalu SS (2009) Bleaching powder: a versatile adsorbent for the removal of fluoride from aqueous solution. Desalination 243:22–31

Karadag D, Tok S, Akgul E, Turan M, Ozturk M, Demir A (2008) Ammonium removal from sanitary landfill leachate using natural Gördes clinoptilolite. J Hazard Mater 153:60–66

Kariyanna H (1987) Geological and geochemical environment and causes of fluorosis—possible treatment—a review. In: Proceedings seminar on role of earth sciences in environment, Bombay, pp 113–122

Ma Y, Wang SG, Fan M, Gong WX, Gao BY (2009) Characteristics and defluoridation performance of granular activated carbons coated with manganese oxides. J Hazard Mater 168:1140–1146

Ma Y, Shi F, Zheng X, Ma J, Gao C (2011) Removal of fluoride from aqueous solution using granular acid-treated bentonite (GHB): batch and column studies. J Hazard Mater 185:1073–1080
Maheshwari RC (2006) Fluoride in drinking water and its removal. J Hazard Mater 137:456–463
Mameri N, Yeddou AR, Lounici H, Grib H, Belhocine D, Bariou B (1998) Defluoridation of septentrional Sahara water of North Africa by electrocoagulation process using bipolar aluminium electrodes. Water Res 32:1604–1610
Martyn CN, Osmond C, Edwardson JA, Barker DJ, Harris EC, Lacey RF (1989) Geographical relation between Alzheimer's disease and aluminium in drinking water. The Lancet 333:61–62
Mckee RH, Johnston WS (1934) Removal of flourides from drinking water. Ind Eng Chem 26:849–851
Miller RG, Kopfler FC, Kelty KC, Stobler JA, Ulmer NS (1984) The occurrence of aluminum in drinking water. J Am Water Works Assoc 84–91
Min Y, Hashimoto T, Hoshi N, Myoga H (1999) Fluoride removal in a fixed bed packed with granular calcite. Water Res 33:3395–3402
Mohan D, Sharma R, Singh VK, Steele P, Pittman CU, Jr CU (2012) Fluoride removal from water using biochar, a green waste, low-cost adsorbent: equilibrium uptake and sorption dynamics modeling. Ind Eng Chem Res 51:900–914
Mohan D, Kumar S, Srivastava A (2014) Fluoride removal from ground water using magnetic and nonmagnetic corn stoverbiochars, 2014. J Ecol Eng 73:798–808
Mohapatra M, Rout K, Gupta S, Singh P, Anand S, Mishra B (2010) Facile synthesis of additive-assisted nano goethite powder and its application for fluoride remediation. J Nanopart Res 12:681–686
Muthukumaran K, Balasubramanian N, Ramakrishna TV (1995) Removal of fluoride by chemically activated carbon. Ind J Environ Protec 15:514–517
Nava CD, Rios MS, Olguin MT (2003) Sorption of fluoride ions from aqueous solutions and well drinking water by thermally heated hydrocalcite. Sep Sci Technol 38(1):131–147
Ndé-Tchoupé AI, Crane RA, Mwakabona HT, Noubactep C, Njau KN (2015) Technologies for decentralized fluoride removal: testing metallic iron-based filters. Water 7:6750–6774
Ndiaye PI, Moulin P, Dominguez L, Millet JC, Charbit F (2005) Removal of fluoride from electronic industrial effluent by RO membrane separation. Desalination 173:25–32
Padmavathy S, Amali J, Raja RE, Prabavathi N, Kavitha B (2003) A study of fluoride level in potable water of Salem district and an attempt for defluoridation with lignite. Ind J Environ Protec 23:1244–1247
Pang XM, Zhang L, Sun SH, Liu T, Gao XH (2007) Effects of metal modifications of Y zeolites on sulfur reduction performance influid catalytic cracking process. Catal Today 125:173–177
Ramos RL, Ovalle-Turrubiartes J, Sanchez-Castillo MA (1999) Adsorption of fluoride from aqueous solution on aluminium-impregnated carbon. Carbon 37:609–617
Rojas-Mayorga CK, Silvestre-Albero J, Aguayo-Villarreal IA, Mendoza-Castillo DI, Bonilla-Petriciolet A (2015) A new synthesis route for bone chars using CO_2 atmosphere and their application as fluoride adsorbents. Micropor Mesopor Mater 209:38–44
Rongshu W, Li H, Na P, Ying W (1995) Study of a new adsorbent for fluoride removal from waters. Water Qual Res J Can 30:81–88
Sarkar M, Banerjee A, Pramanick PP, Sarkar AR (2006) Use of laterite for the removal of fluoride from contaminated drinking water. J Colloid Interface Sci 302:432–441
Schneiter RW, Middlebrooks EJ (1983) Arsenic and fluoride removal from groundwater by reverse osmosis. Environ Int 9:289–291
Sorg TJ (1978) Treatment technology to meet the interim primary drinking water regulations for inorganics. J. Am Water Works Assoc 70:105–111
Sovinelli EA, Black AP (1958) Defluoridation of water with activated alumina. J Am Water Works Assoc 50(1):33–44
Srimurali M, Pragathi A, Karthikeyan JA (1998) Study on removal of fluoride from drinking water by adsorption onto low cost materials. Environ Pollut 99:285–289

Sun Y, Fang Q, Dong J, Cheng X, Xu J (2011) Removal of fluoride from drinking water by natural stilbite zeolite modified with Fe (III). Desalination 277(1):121–127
Thergaonkar VP, Nawalakhe WG (1971) Activated magnesia for fluoride removal. Ind J Environ Health 16:241–243
Tripathy SS, Bersillon JL, Gopal K (2006) Removal of fluoride from drinking water by adsorption onto alum-impregnated activated alumina. Sep Purif Technol 50:310–317
Wang Y, Reardon EJ (2001) Activation and regeneration of a soil sorbent for defluoridation of drinking water. Appl Geochem 16:531–539
Wasay SA, Haran MJ, Tokunaga S (1996) Adsorption of fluoride, phosphate, and arsenate ions on lanthanum-impregnated silica gel. Water Environ Res 68:295–300
WHO (World Health Organization) (1984) Guidelines for drinking water quality (vol. II): health criteria and supporting information. World Health Organization, Geneva, Switzerland

Chapter 13
Applications of Remote Sensing and GIS in Water Quality Monitoring and Remediation: A State-of-the-Art Review

Meenu Ramadas and Alok Kumar Samantaray

Abstract Recent advancements in the field of remote sensing and geographic information systems (GIS) have made it possible to conduct large-scale water remediation studies. Using improved spectral and spatial resolution sensors and geospatial modeling techniques, water quality parameters such as chlorophyll-a, algae bloom, turbidity, suspended sediments, and mineral content in water bodies including groundwater are being monitored at low cost and with greater accuracy. Integration of these technologies with field monitoring have successfully aided in identification of contamination zones and sources of contamination, and for developing strategies for remediation. High-resolution mapping of contamination zones will further help in allocating remediation efforts to the critically affected areas. This chapter investigates the status of ongoing research in the domain of remote sensing and GIS for water quality monitoring and management, and remediation of water resources.

Keywords Remote sensing · Water quality · Water remediation
Contamination mapping · GIS

13.1 Introduction

Water quality is known to determine the health of society and ecosystems, and any deterioration in its quality presents a risk to human health, food productivity, and biodiversity (Ritchie and Cooper 2001; Tebbs et al. 2013). Water bodies and supply

M. Ramadas (✉) · A. K. Samantaray
School of Infrastructure, IIT Bhubaneswar, Khordha 752050, Odisha, India
e-mail: meenu@iitbbs.ac.in

A. K. Samantaray
e-mail: aks14@iitbbs.ac.in

S. Bhattacharya et al. (eds.), *Water Remediation*, Energy, Environment, and Sustainability, https://doi.org/10.1007/978-981-10-7551-3_13

systems that cater to the water demands of the society are often contaminated by chemical/oil spills, leaks of industrial tanks and pipelines, chemicals from pesticide and fertilizer, and leachates to underground water aquifers from landfills, and septic systems. The common contaminants include arsenic, cadmium, chloramine, chromium, copper, fluoride, bacteria and viruses, lead, nitrates/nitrites, mercury, perchlorate, radium, selenium, silver, and uranium. Besides these, there are the emerging contaminants: from pharmaceuticals, personal care products (PCPs), and endocrine disrupting compounds (EDCs), whose effect is not fully understood yet (Petrovic et al. 2004).

Water remediation is the process that deals with the control of water pollution as well as removal of the different types of contaminants from sources of water supply. The process has multiple objectives, ranging from water purification to the safe disposal of residues in an environment-friendly manner. Therefore, several in situ and ex situ strategies (biological, physical, and chemical treatment methods) are employed in remediation of water bodies until the water meets the purity standards set by State and National/Federal agencies. Various parameters such as chlorophyll-a concentration (chl-a), turbidity, salinity, Secchi disk depth (SDD), temperature, pH (acidity/alkalinity), total dissolved solids (TDS), total suspended solids (TSS), total organic carbon (TOC), total hardness (TH), and concentrations of potassium (K), phosphorus (P), and nitrogen (N) are evaluated in water quality studies (Kulkarni 2011; Gholizadeh et al. 2016). The contaminant concentrations set by the standards vary for drinking water and water to be used for agricultural purposes.

Field monitoring by collecting water samples from water bodies at regular intervals has been the primary source of information on water quality for monitoring purposes. However, the time and costs associated with this exercise is a concern while conducting large-scale spatiotemporal studies on water quality of freshwater systems.

There are natural and anthropogenic activities or events that alter the natural condition of the water with time, such as wildfires, thunderstorms and heavy rainfall, changes in vegetative cover or forest clearing, cultivation, and fertilizer application. Fortunately, technological advancements in the last few decades have provided solutions through remote sensing and geographical information systems (GIS) platforms, to address several natural resources-related issues. The changing composition and properties of coastal waters and inland water bodies, sources of pollution, and transport of contaminants can be measured through remote sensing with synoptic scale coverage (Markogianni et al. 2014; Kallio et al. 2015). Satellites are helpful in oil spill disaster response by providing monitoring/assessment for remote locations (Brekke and Solberg 2005; Jha et al. 2008). Several authors including Griffith (2002), Ritchie et al. (2003), Shen et al. (2012), Kumar and Reshmidevi (2013), Palmer et al. (2015), and Gholizadeh et al. (2016) have extensively reviewed the remote sensing approach to water quality monitoring.

Remote sensing-based techniques have been found to assist with water quality monitoring and management; however, geographical information systems (GIS) is what makes it possible to seamlessly link water quality monitoring in space and time with remediation strategies at reduced costs. GIS has also facilitated the collection of data, and communication of water quality hazards in real time, through display maps. This chapter reviews the applications of remote sensing and GIS in monitoring of water quality and remediation of water resources, both surface and ground water. The rest of the chapter is organized as follows: Sect. 13.2 describes the scope of remote sensing and GIS in water quality monitoring and remediation, Sects. 13.3 and 13.4 discuss the various applications of these techniques in surface and groundwater quality studies, respectively. Section 13.5 addresses the future scope of remote sensing and GIS in water quality studies, and Sect. 13.6 summarizes the findings based on the review.

13.2 Remote Sensing and Geographical Information Systems (GIS)

13.2.1 Scope of Remote Sensing

Remote sensing works on the principle that different constituents of water reflect or emit different amount of energy in different wavelength ranges (or bands) of the electromagnetic spectrum that falls on them. Pollutants in water may change the reflectance of water based on their structural, chemical, and physical properties (Jerlov 1976; Kirk 1983). Sensors aboard different remote sensing satellites and aerial vehicles are able to record the reflected/emitted energy, and that is processed to form an image that provides valuable information on the extent of contamination of water bodies and concentration of contaminants.

The process of remote sensing requires a steady source of energy that can emit electromagnetic radiations in the wavelength regions to be sensed by the sensors. Direct sources such as radar and indirect sources such as the Sun are examples of energy sources. Remote sensing sensors that use the natural sources such as the Sun are known as passive sensors, whereas sensors that utilize their own source of energy are active sensors (Lillesand and Kiefer 1994). The images retrieved from remote sensing vary in quality based on different sensor characteristics such as their spatial resolution, spectral resolution, radiometric resolution, and temporal resolution. When they operate in the visible, near-infrared (NIR), middle infrared (MIR), and shortwave infrared (SWIR) portion of the electromagnetic spectrum (300–3000 nm), it is known as optical remote sensing, those operating in thermal range of electromagnetic spectrum (3000–5000 and 8000–14,000 nm) constitutes thermal remote sensing, whereas microwave remote sensing employs microwaves in the wavelength range of 1 mm to 1 m of electromagnetic spectrum.

Satellite images are freely as well as commercially available at various spatial resolutions, namely Moderate Resolution Imaging Spectroradiometer (MODIS), Medium Resolution Imaging Spectrometer (MERIS), WorldView-2, WorldView-3, Landsat 5 TM, Landsat 7 ETM, Landsat 8, GeoEye, NOAA AVHRR, IKONOS, Nimbus 7, Formosat, Cartosat, Advanced Space borne Thermal Emission and

Reflection Radiation (ASTER), Indian Remote Sensing (IRS) LISS-3 and P6, QuickBird, HICO, SeaWIFS, and SPOT HRV. There are multispectral images that have multiple image layers, each layer representing image acquired at a particular wavelength band. For instance, SPOT HRV uses radiations in three wavelength bands: the green (500–590 nm), red (610–680 nm), and NIR (790–890 nm) bands, and IKONOS uses four bands in the spectrum: blue, green, red and NIR, while a Landsat TM has seven bands: blue, green, red, NIR bands, two SWIR bands, and a thermal IR band. In addition to remote sensing satellites, hyperspectral airborne missions are conducted to assess water quality. AVIRIS, HYDICE, Hymap, MVIS, DAIS, CASI 1500 are few examples of airborne sensors used in water quality studies. Microwave radiometers (MWR) measure microwave emissions from topographic features to assess sea surface temperature (SST) and water quality parameters such as salinity. AQUA, TRMM, Nimbus 5, Nimbus 7, and SEASAT satellites work in the microwave range of the electromagnetic spectrum.

As mentioned earlier, the spectral signatures collected by sensors provide information on contaminants and their concentration. Spectral reflectance of water is low in all parts of the electromagnetic spectrum. However, reflectance increases in the visible portion due to different materials suspended in water (Lillesand and Kiefer 1994) and helps in distinguishing between clear water and the sediment-laden water. With increased algae concentration, the reflectance will increase across most of the wavelengths. Similarly, plant canopy light reflectance property could be used to detect the presence of weeds in water bodies. Water hyacinth has higher NIR reflectance than other plant species and water. Hydrilla has lower NIR reflectance than other plant species but higher NIR reflectance than water. Water hyacinth and hydrilla are similarly distinguishable in color infrared (CIR) imagery; they have bright orange-red and reddish-brown image responses, respectively (Everitt et al. 1999).

13.2.2 Scope of GIS

Geographical information systems (GISs) are a powerful tool in natural resource-related studies. According to the ESRI (1990), it encompasses use of computer hardware, software, geographic data, and personnel designed to work with all forms of geographically referenced information. This integration of topographical features with physical parameters has enabled decision making and has been used extensively to monitor changes in land use–land cover, changes in population density, housing, and flood zone mapping. GIS database works with different thematic maps or layers that carry different types of information such as elevation, soil type, land use, and population density, to name a few. The information in any two layers may be combined and transformed into a new layer for research studies. There are primarily two data types in GIS: (i) raster data that stores geographic information in pixels, and (ii) vector data that stores geographic features in the form of points, lines, and areas. The spatial resolution of a raster is defined by the size of each pixel, and it is suitable for attributes that exhibit a high degree of variation in space.

GIS is highly useful for making decisions based on spatial data that have been collected from field sampling and remote sensing. Using hand-held global positioning system (GPS) units, coordinates of the field locations where water quality samples are collected can be recorded, and then entered into the GIS database. Information from satellite images, aerial photographs, soil maps, topographic maps, and local survey maps can be incorporated in GIS. In water quality management and remediation studies, GIS may be used to investigate, for instance, the spatial relationship between contaminant concentrations and topography or land use–land cover or soil type (Srinivasa Rao and Jugran 2003; Everitt et al. 1999). Similarly, GIS allows hazard zone mapping for pollution in water bodies and agricultural lands adjacent to water bodies. GIS enables monitoring changes that have taken place over time within a watershed/area and linking changes in water quality with these changes (Oikonomidis et al. 2015).

13.3 Surface Water Quality

13.3.1 Surface Water Quality Monitoring

Over the last few decades, remote sensing and GIS have been successfully utilized in surface water quality monitoring studies. For studies on inland and coastal water bodies, limnologists have been using satellite images from ERTS-1/Landsat 1 launched since 1972 (Bukata 2013). An integrated research framework involving spatial features, data collected, and computational database is an essential component for these studies (Ritchie et al. 2003). Sensors mounted on satellites and airplanes measure the amount of radiation at various wavelengths reflected from the surface of water bodies that are then translated to different water quality indicators. With advancements in the field of remote sensing, parameters such as TSS, chl-a, turbidity, and SDD are monitored fairly well.

Aerial photographs collected during remote sensing are analog images, whereas satellite imagery is digital, i.e., they are composed of pixels. The intensity of a pixel is recorded as a digital number (DN), taking an average value for the whole pixel area. The exact range of DN of a sensor depends on its radiometric resolution (0–63, 0–255, or 0–2047). The satellite imagery undergoes processing, i.e., conversion of raw digital numbers into a meaningful spectral radiance value. Using calibration curve of DN to radiance prepared by the satellite operators, DN is converted into spectral radiance L as shown in Eq. (13.1). For instance, conversion of 8 bit (DN in the range 0–255) Landsat imagery requires value of gain and bias of the sensor in each band:

$$L = \text{Bias} + (\text{Gain} * \text{DN}) \tag{13.1}$$

where L denotes the spectral radiance. While gradient of the calibration gives gain, bias is obtained as the spectral radiance of the sensor for a DN value of zero. The spectral radiance values of different features are then converted to planetary reflectance as shown in Eq. (13.2) so that they are comparable across sensors and time.

$$P = \left(\pi * L * d^2\right) / (\text{Irradiance} * \sin(\pi * \text{sun elevation angle}/180)) \tag{13.2}$$

where P is reflectance, i.e., the ratio of radiance to irradiance, and it is a standardized measure. The Earth–Sun distance (d) and sun elevation angle data vary with the date of image acquisition. Exo-atmospheric/planetary solar irradiances are specifically available for different bands of different satellite sensors. Atmospheric correction algorithms are then applied so that attenuation effects caused by aerosol scattering and absorption by water vapor, ozone, oxygen, and carbon dioxide are removed from surface reflectance retrievals (Martins et al. 2017).

In the early 1970s, studies had developed statistical relationships between spectral properties from sensors and water quality parameters (Ritchie et al. 1974, 2003). These empirical equations were of the forms given by Eqs. (13.3) and (13.4):

$$Y = A + B * X \tag{13.3}$$

$$Y = A * B^X \tag{13.4}$$

where Y is the remotely sensed radiance/reflectance, and X is a water quality parameter such as chl-a or suspended sediments. A and B in this equation are empirically derived factors. A simple example of application of reflectance values in water quality studies is in the estimation of normalized difference vegetation index (NDVI). The NDVI algorithm is given by Eq. (13.5); it utilizes red and NIR reflectance values:

$$\text{NDVI} = (\text{NIR} - \text{Red}) / (\text{NIR} + \text{Red}) \tag{13.5}$$

Over water bodies, negative NDVI values (close to -1) indicate the presence of water whereas positive NDVI values indicate presence of contaminants (Holme et al. 1987; Akbar et al. 2014). Previous studies have suggested that watershed vegetation cover reflects the runoff characteristics and water quality (Munn 1988; Kelly and Harwell 1990; Jones et al. 1997). Study by Griffith et al. (2002) and Griffith (2002) concluded that the NDVI performed better than landscape pattern metrics or land use/cover proportion information in explaining variation in water quality parameters. For instance, suspended solids and nitrate concentrations were found to be correlated with NDVI values at watershed scale (Chu et al. 2013). Previous studies have developed quantitative relationships between water-leaving radiance received by satellite, and in situ measured suspended particles concentration data (Curran and Novo 1988). Similarly, chlorophyll content can be estimated using Eq. (13.6) (Harding et al. 1995):

$$\log_{10}[\text{Chlorophyll}] = a + b * (-\log_{10} G) \qquad (13.6)$$

where a and b are empirical constants obtained from field measurements, G is $[(R_2)^2/(R_1 * R_3)]$ and R_1 is radiance at 460 nm, R_2 is radiance at 490 nm, and R_3 is radiance at 520 nm.

The algorithms used for water quality monitoring can be summarized as empirical models (Tassan 1993), physical models based on radiative transfer theory (Doerffer and Fischer 1994), and semi-analytical models that combine physics with statistical methods (Zhang et al. 2016). A review by Odermatt et al. (2012) details the suite of band arithmetic and spectral inversion algorithms used to estimate optical properties of water. Kaba et al. (2014) list few of the studies that have reported correlations between MODIS measured reflectance and turbidity, total suspended matter (TSM), suspended sediment concentration (SSC), and/or TSS. In a recent study, Tebbs et al. (2013) developed an algorithm for monitoring of chl-a that indicates presence of cyanobacterial biomass, by utilizing its strong correlation with Landsat reflectance ratio, R835/R660.

Landsat satellite imagery has been used extensively by scientists for water quality monitoring. Landsat is the National Aeronautics and Space Administration's (NASA) Earth observation program that provides imagery since 1972. It is equipped with MSS (Multispectral Scanner), TM (Thematic Mapper), and ETM+ (Enhanced Thematic Mapper Plus), and covers majority of the Earth's landmass. As early as the 1980s, the imagery from this mission was put to use. Nelson et al. (2003) utilized Landsat 7 Enhanced Thematic Mapper Plus (ETM+) imagery of the lower peninsula of Michigan, USA, for monitoring the lake water clarity. Zhang et al. (2003) had used a combination of Landsat TM bands and ERS-2 SAR-derived backscattering coefficients to estimate parameters such as turbidity, SDD, and suspended sediment concentration for the Gulf of Finland region. In a series of studies conducted by Hellweger et al. (2004), estuarine turbidity of Hudson River and the eutrophic extent of East River in New York were mapped, as the SDD and chl-a had a good correlation with Landsat TM red reflectance and with the ratio of Landsat TM green to red reflectance (TM2/TM3), respectively. Alparslan et al. (2007) had mapped chl-a, suspended solids, turbidity, and total phosphate at the Ömerli Dam site using Landsat ETM satellite image. This was achieved by exploring the relationship between pixel reflectance bands 1–4, and the water quality parameters. The parameters chl-a, suspended matter, and turbidity change the watercolor. Chl-a, in fact, have high reflectance in green and low reflectance in red bands, and similarly, TSS also correlates with ratios, logarithmic transformations and likewise combinations of Landsat TM spectral bands (Kulkarni 2011). In fact, coastal salinity studies on estuarine regions have also been benefitted by the remote sensing satellites. Survey of spatial variation of turbidity and TSS at Pearl River Estuary, China, was carried out using EO-1 ALI satellite imagery. The Advanced Land Imager (ALI) instrument on NASA EO-1 was meant for economizing future Landsat satellites, in terms of mass, size, power consumption, and cost, and improved instrument sensitivity and image resolution (Digenis 2005).

Kilham et al. (2012) developed models for measuring the suspended sediment concentration (SSC) in Feather River, California, USA, especially under flood conditions, using Landsat remote sensing images. They applied the sediment mineralogy and particle-size distribution and calibrated above-water reflectance field spectra with known SSC values in their algorithm. Kutser (2012) had evaluated performance of Landsat satellite images for studying lake carbon content by mapping colored dissolved organic matter (CDOM) changes in Sweden.

Similarly, Choubey and Subramanian (1990) had explored the capabilities of IRS 1A-LISS 1 for measurement of suspended solids in Tawa Reservoir, India. They concluded that the grain size and mineral composition of sediments affect reflected radiance in visible wavelength bands 1, 2, and 3, especially at lower concentration levels. Among other satellites utilized, Matthews et al. (2010) had investigated on empirical algorithms for estimating chl-a, TSS, absorption by CDOM and SDD of Zeekoevlei lake, South Africa, using the imagery from Medium Resolution Imaging Spectrometer (MERIS) onboard the European Space Agency's (ESA) Envisat. Besides these, MERIS satellites were used previously for inland water quality monitoring in moderately sized water bodies, for detecting chl-a (Giardino et al. 2005), and phycocyanin (PC, Simis et al. 2005). Recent technological advancements and sensors such as Airborne Visible/Infrared Imaging Spectrometer (AVIRIS) and Airborne Imaging Spectrometer for Applications (AISA) have enabled remote sensing to assist in oil spill response and monitoring. Figure 13.1 shows the images of oil spill along Louisiana Coast, USA, in true color and false color obtained from the Multi-angle Imaging SpectroRadiometer (MISR) instrument onboard NASA's Terra spacecraft. The extent of the spillover a 100 km by 100 km region could thus be visualized using remote sensing imagery. In Fig. 13.1, in the true color image on the right-hand side, the oil film is visible in blue-black shade, near the Mississippi River Delta in Louisiana (NASA 2010). In the suite of high-resolution spaceborne sensors, the QuickBird satellite of DigitalGlobe offers highly accurate, commercial high-resolution imagery. QuickBird satellite imagery was used by Yüzügüllü and Aksoy (2011) for identification of eutrophic zones in the Lake Eymir region, wherein the authors developed an ANN model to derive SDD values using the red, blue, green, and NIR bands. SDD can assess the eutrophic characteristics of the lake. Akbar et al. (2014) had established a Landsat 5 TM-based water quality index that relates the planetary reflectance (utilizing four spectral bands—blue, green, red, and NIR) to monthly ground measured data. Their index had five classes of water quality (excellent, good, fair, marginal, and poor), and six classes of turbidity, namely 0–10, 10–20, 20–30, 30–40, 40–50, and >50 Nephelometric Turbidity Unit (NTU). A modified radiative transfer model relating the surface directional remote sensing reflectance (calculated from remote sensing observations after applying atmospheric correction), with the properties of suspended solids in the water column was developed by Volpe et al. (2011) to analyze water quality and sediment problems in the Venice lagoon. The use of multispectral satellite sensors such as Landsat TM5, Landsat ETM 7, ASTER, and ALOS AVNIR was beneficial in this study. Singh et al. (2013) had developed a water quality indicator using MODIS remote sensing

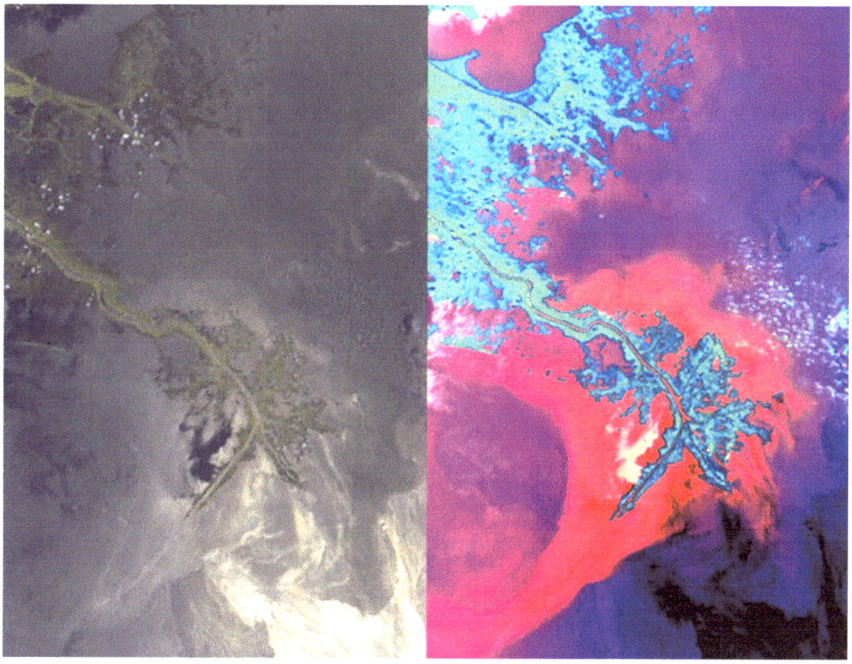

Fig. 13.1 Oil spill images captured by the Multi-angle Imaging SpectroRadiometer (MISR) instrument onboard NASA's Terra spacecraft. Left-hand image is in near-true color, but vegetation appearance has been enhanced, while the right-hand image is in false color, oil spill represented in blue-black color. *Source* NASA (2010), *Image Credit* NASA/GSFC/LaRC/JPL, MISR Team

data for the state of Wisconsin, USA. Surface reflectance in the blue, red, and IR bands from MODIS was combined to estimate the NDVI and Enhanced Vegetation Index (EVI) that were then used to predict the nitrate and dissolved phosphorus concentrations in stream water. WorldView-2 satellite imagery was utilized by El Saadi et al. (2014) to obtain water quality parameters for the Rosetta River Nile Branch, Egypt. However, only the optical parameters such as TSS and chl-a could be successfully retrieved from the multispectral imagery. A recent study by Abdelmalik (2016) used the ASTER data to monitor water quality parameters such as temperature, turbidity, pH, salinity, TDS, conductivity, TOC, and ortho-phosphorus. Satellite remote sensing could also be used to determine nutrient distributions in lakes with reasonable confidence (El Saadi et al. 2014; Chen et al. 2009; Volpe et al. 2011). Images from Hyperspectral Imager for the Coastal Ocean (HICO) on the International Space Station (ISS) were employed to derive chl-a, CDOM absorption, and turbidity in northern Florida. HICO is the first space-based, maritime hyperspectral imaging instrument that has capabilities to monitor water quality in the coastal regions (Keith et al. 2014).

Using airborne visible/near-infrared imaging spectroscopy, spatial coverage of water quality studies conducted has improved with advantages such as lower costs and higher resolution, compared to satellite imagery. The technique of airborne scanning was used in Western Australia by Jupp et al. (1994) for studying the distribution of algal species, cyanobacteria, and associated water parameters (such as turbidity). The digital Compact Airborne Spectrographic Imager (CASI) employed for this purpose had hyperspectral sensors that collect several hundred spectral bands of data at a high spatial resolution and was mounted to a light aircraft. Harding et al. (1995) had combined aircraft and field-based measurements to measure and map chlorophyll content in the Chesapeake Bay. An airborne sensor, portable remote imaging spectrometer (PRISM) was used to monitor water quality parameters: turbidity, DOC and chl-a concentrations, and even the surface methylmercury concentrations at high spatial resolution (2.6 m grid) in the San Francisco Bay–Delta Estuary wetlands (Fichot et al. 2015).

13.3.2 Surface Water Quality Remediation

Similar to water quality monitoring studies, remote sensing has accomplished milestones in surface water remediation. GIS is utilized to process spatial data to study the risk of surface water pollution (Tiktak et al. 1996; Sinkevich et al. 2005; Lim and Choi 2015). There are powerful spatial analysis softwares such as ArcGIS to process spatial data of water quality monitoring and maps (ESRI 1990).

Identification of critical hot spots is a crucial step in remediation studies. This is more relevant in cases of non-point source pollution (NPS) in watersheds as the spatial extent of contamination into surface and groundwater is fairly uncertain. NPS pollutants (sediments, fertilizers, pesticides, salts, etc.) are diffuse in nature and cannot be traced to a point location, and it is not feasible to determine the areal extent of their contamination. Recent advances have given rise to watershed-scale distributed modeling tools in GIS environment to model nutrient and pollutant movement, such as the Chemicals, Runoff, and Erosion from Agricultural Management Systems (CREAMS, Frere et al. 1980), Soil and Water Assessment Tool (SWAT, Arnold et al. 1996), SPAtially Referenced Regressions On Watershed attributes (SPARROW, Alexander et al. 2007; Robertson et al. 2009). Spatially Explicit Delivery MODel (SEDMOD, Fraser et al. 1998) with GIS support could estimate eroded sediment or NPS pollutant load transported from each cell in the watershed to a stream channel. The authors recommend SEDMOD for application over large areas to prioritize sites for NPS pollution. In order to control sediment loads and to assess performance of conservation practices in reducing sediment and contaminant loadings in the Great Lakes Basin, a GIS-based coupled erosion-sediment delivery model was used by Ouyang et al. (2005). SPARROW model is used to estimate pollutant sources and contaminant transport in watersheds and surface waters. Studies using SPARROW have successfully assessed the sources of nutrient loadings in streams, and utilized it for nutrient reduction in watersheds. Robertson et al. (2009) had investigated application of this model to identify areas with "high priority" in the Mississippi–Atchafalaya River basin based on delivery of nutrient yields to the Gulf of Mexico. Extreme water degradation instances such as hypoxia could be prevented by knowing the delivered nutrient yields and by reducing nutrient loadings from these areas to downstream streams and reservoirs.

Hunter et al. (2010) suggest that data from remote sensing satellites could contribute significantly to the remediation efforts of lakes in case of cyanobacterial blooms and presence of toxins (cyanotoxins). They had investigated on the algorithms for the retrieval of chl-a and phycocyanin (C-PC) from data acquired by the CASI-2 and the Airborne Imaging Spectrometer for Applications (AISA) Eagle Sensor. By combining hyperspectral remote sensing, GPS, and GIS technologies, mapping of weed and invasive species infestations over large and inaccessible waterways could be performed, to aid in restoration of freshwater, estuarine, and floodplain habitats. Airborne hyperspectral imager HyMap has 128 bands in the visible and NIR and SWIR and was employed by Hestir et al. (2008) for studies on California's Sacramento–San Joaquin River Delta. The HyMap images were converted to apparent surface reflectance for detection of non-native species such as water hyacinth and submerged aquatic vegetation, and using GIS software, mapping of their distribution in the waterways was performed. Shen et al. (2012) have reviewed the different satellite sensors, techniques, and algorithms that have been employed for detecting harmful algal blooms, in one of their studies.

Active airborne and spaceborne synthetic-aperture radar (SAR) systems can provide rapid response products in case of oil and contaminant spills (Brekke and Solberg 2005; Jha et al. 2008). Using remote sensing, maps of oil thickness can be developed that are utilized in the post-spill recovery efforts and impact monitoring using hyperspectral imagery and airborne SAR imagery, respectively (Leifer et al. 2012). Similarly, by closely monitoring spatial and temporal changes of vegetation growing on landfills, contaminant leaks could be detected (Jensen et al. 2009). Contaminants such as copper or zinc, and water movement in hazardous waste sites affect the leaf area index, biomass, etc., of vegetation, which in turn influence the spectral reflectance pattern. In this context, containment and remediation of surface water bodies from potential water quality deterioration could be easily accomplished.

Linking land use–land cover changes with changes in surface water quality, Liu et al. (2009) utilized statistical methods and GIS to identify contaminated watersheds in Wisconsin, USA. Their study performed map overlay of agricultural and urban land use with distribution maps of phosphorus, lead, and fecal coliform to identify contaminates zones. Sridhar et al. (2009) had quantified and mapped total phosphorous concentration in surface soils using Landsat imagery of bare soil fields. The authors analyzed Ba, Cd, Cu, S, and P element concentrations in the surface soils using Landsat 5 TM imagery. Yenilmez et al. (2011) demonstrated how GIS could be utilized for locating areas in a mine site that are more likely to have high pollutant and trace elements (Mn, Cr, Fe, S) concentrations, thereby bringing downtime and sampling costs. Using the ArcGIS software, the spatial distribution of pollutants, topographical features, surface runoff pathways, and locations of contamination sources at the mine site was studied together to derive the pollution hot spots. Under an international consortium, the Hotspot Ecosystem Research and Man's Impact On European Seas (HERMIONE) project was implemented during 2009–2012 to investigate hot spots in the European region.

They focused largely on the underwater ecosystems, and study areas selected for the investigations are shown in Fig. 13.2 (NOC 2009). The use of GIS makes it possible to develop spatial maps of contamination hot spots in vulnerable ecosystems. Similarly, with the help of GIS, data collected from field investigations of water quality can be incorporated in maps, to explain the impacts of human intervention. For instance, Fig. 13.3 shows the results of a stream water quality sampling study conducted in the states: Alabama, Georgia, North Carolina, South Carolina, and Virginia in the USA by the United States Geological Survey (USGS). The focus study areas as well as the cumulative concentrations of pharmaceuticals that were deteriorating water quality are conveyed by using the map (USGS 2016).

Under regulations such as the Clean Water Act of the USA, water bodies contaminated by pollutants are identified so that they could be cleaned up using various

Fig. 13.2 Study areas chosen by the Hotspot Ecosystem Research and Man's Impact On European Seas (HERMIONE) project that are vulnerable to the impacts of climate change and human activities. *Source* National Oceanography Centre, Southampton, UK

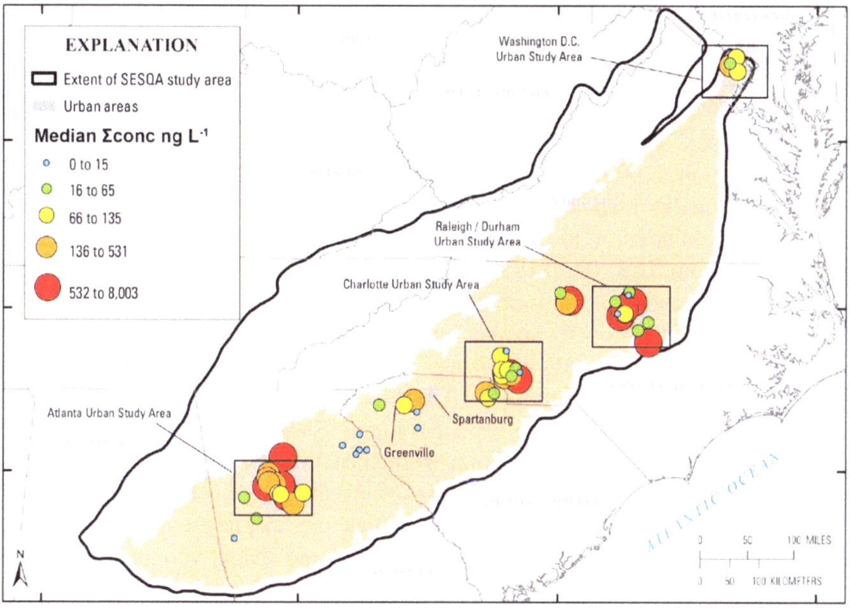

Fig. 13.3 GIS-based mapping of pharmaceutical chemical contamination in the USA. Colored circles of varying size indicate various ranges of concentrations of chemicals in streams in the different urban study areas in the region represented by black boxes (image courtesy of the USGS)

strategies. Such efforts must be adopted by national and state governments in every country, and remediation can be monitored using remote sensing and GIS without much expense. Total maximum daily loads (sum of watershed point sources, non-point sources, and a margin of error) that will not impair the water body need to be recalculated for major pollutants such as suspended sediments, pathogens, nutrients, metals, dissolved organic matter, pesticides, algae, plants, temperature, and oils (Ritchie and Cooper 2001). The advanced remote sensing and GIS techniques will also help in evaluating the effectiveness of the efforts that are adopted for mitigation.

13.4 Groundwater Quality

13.4.1 Groundwater Quality Monitoring

Several groundwater quality studies involving remote sensing and GIS were carried out in the last two decades. Although remote sensing is best suited for applications involving surface parameters and variables, with recent advances in the sensors deployed and physical hydrological models, it has been possible to achieve greater strides in groundwater studies. In one of the earlier studies, Maidment (1996) investigated the capabilities of GIS and hydrologic models to model the transport of agricultural pollutants in the vadose zone. His study attributed the presence of nitrate in Texas groundwaters to the hydrogeologic topographical features. A study by Ahn and Chon (1999) had linked groundwater quality in Seoul, South Korea, with topography, geology, land use, and pollution sources, using GIS. Srinivasa Rao and Jugran (2003) used remote sensing and GIS technology to prepare groundwater quality maps of TDS, hardness, incrustation problem, and magnesium concentration, in addition to mapping of groundwater potential. They had used the image data from IRS ID (LISS III) and Landsat 5 TM in conjunction with Integrated Land and Water Information System (ILWIS) software. ILWIS is a free remote sensing and GIS software that integrates remote sensing and GIS platforms, and facilitates import/export, digitizing, editing, analysis, and display of spatial data, to yield maps. Alternatives to ILWIS include the popular ArcGIS, QGIS, GRASS, and ERDAS Imagine among others. Basavarajappa and Manjunatha (2015) studied rock–water interaction, residence time of water in aquifers, and human and natural activities, in the context of groundwater quality. The parameters included fluoride, nitrate, carbonate, chloride, calcium, magnesium, sodium, sulfate, iron, potassium, TDS, pH, and TH. They mapped the spatial variation of groundwater quality in Precambrian hard rock terrain of Chitradurga District using IRS 1D, LISS III, PAN, and PAN+LISS III images and GIS software.

Numerous studies have utilized the changes in land use and land cover to address the issue of water quality deterioration (Asadi et al. 2007; Singh et al. 2011). Babikar et al. (2004) could correlate nitrate concentrations in groundwater with land use vegetable cultivation, in their GIS-based analysis of groundwater contamination by nitrate. Babiker et al. (2005)'s study showed that the eastern part of the Kakamigahara aquifer in Japan was vulnerable to groundwater pollution due to intensive vegetable cultivation. Nas and Berktay (2010) have used geostatistics and ArcGIS to produce spatial distribution maps of groundwater quality parameters such as pH, electrical conductivity, chloride, hardness, and nitrate^{-} concentrations. These studies used kriging for spatial analysis of water quality parameters. Geostatistical methods for spatial interpolation such as inverse distance weighting, ordinary kriging, and universal kriging method are inevitable in GIS environment for mapping water quality parameters (Mueller et al. 2004). Singh et al. (2011) could identify that weathering and leaching action of feldspars and magnesium calcite as well as industrial contaminants and fertilizers were responsible for deteriorating water quality in their study area, using Landsat and LISS III imagery in a GIS framework. The groundwater present in certain pockets of the region indicated high concentrations of nitrate, manganese, and chromium exceeding permissible limits of WHO. In overall, they rated the resource into very good, good, moderate, poor, very poor, and unfit classes based on groundwater quality index (GWQI) values. Srivastava et al. (2013) had mapped groundwater quality using GIS framework and found out that temporal changes in land use obtained from IRS 1D LISS III satellite imagery could be utilized to understand changes in quality of underground water.

13.4.2 Groundwater Quality Remediation

Groundwater quality monitoring is closely followed by the development of remediation strategies to restore the resource. Coupling of groundwater models with GIS has been popularly used to study NPS pollutants in aquifers. Corwin and Wagenet (1996) suggest that studies need to incorporate geospatial statistics, remote sensing, GIS and flow, and solute transport models to assess NPS pollution remediation strategies. For groundwater remediation, Corwin et al. (1998) coupled GIS with vadose zone transport and mapped the areal distribution of contaminant through the soil profile and solute loading to the ground water. Tiktak et al. (1996) had utilized a one-dimensional pesticide leaching model known as PESTRAS in conjunction with GIS to derive spatial patterns of the pesticide leaching potential in the Netherlands. The nitrate leaching and economic analysis package (NLEAP) model combined with GIS was similarly adopted to model NO_3–N leaching in eastern Colorado (Shaffer et al. 1996). This framework was used to identify the hot spots in shallow alluvial aquifers in the region, and thereafter involved in decision making on best management practices (BMP) to minimize nitrate leaching. Sinkevich et al. (2005) had used preferential flow transport models along with GIS to determine the area in the study region with high risk of contamination by agrochemicals. They relied on NO_3–N concentrations in groundwater to associate with extent of contamination. The Agricultural Pressures and Impacts on European waters (Ag–PIE) model based on GIS was developed by Giupponi and Vladimirova (2006) with an extensive framework to address the concerns associated with diffuse pollution of water resources. This model was built on a DPSIR framework (Driving Force—Pressure—State—Impact—Response) where driving forces include agriculture or any pollution-causing activities that introduce a pressure on the system changing the state (quality of water), resulting in serious changes in the ecosystem and the mitigation efforts to combat the same and reduce pollution. The Ag-PIE model thus provided information helpful to identify the critically affected areas and for policy makers to adopt the best measures or practices for target areas. Oikonomidis et al. (2015) demonstrated that policy makers and decision makers can be benefitted using maps created using remote sensing and GIS while selecting sites suitable for drilling boreholes. Their study was conducted in Tirnavos, Greece, where high nitrate concentrations are present due to the intensive agriculture, and performing an overlay of groundwater potential zones and their nitrate concentrations could identify the polluted zones.

DRASTIC is one of the most popular models that can address groundwater vulnerability to potential contaminants. DRASTIC, developed by the United States Environmental Protection Agency (US EPA), is an overlay and index model that evaluates the potential for groundwater contamination at any location based on geological and hydrological factors (Knox et al. 1993). Depth to groundwater, recharge by rainfall, aquifer type, soil properties, topography, impact of the vadose zone, and the hydraulic conductivity of the aquifer are combined in DRASTIC index. Studies have demonstrated the coupling of DRASTIC and GIS coupling to be useful for risk assessment for groundwater and aquifers (Navulur and Engel

1998; Secunda et al. 1998; Al-Adamat et al. 2003; Thirumalaivasan et al. 2003; Kazakis and Voudouris 2015). Huan et al. (2012) modified the DRASTIC model into the RSIVL model (net recharge R, soil type S, impact of vadose zone I, groundwater velocity V, and land use type L) in order to assess the groundwater vulnerability to nitrate concentration. In all these studies, GIS has facilitated vulnerability mapping and identification of contaminated sites.

13.5 Future of Remote Sensing and GIS Applications in Water Quality Studies

The future of remote sensing applications in water quality studies is promising in several aspects. Water quality monitoring and remediation would benefit from the most recent Sentinel-2 and Landsat 8 imagery (Martins et al. 2017; Toming et al. 2016; Palmer et al. 2015). The new age Sentinel missions were initiated by the ESA under the environmental monitoring Copernicus programme. The Multispectral Imager (MSI) onboard the Sentinel 2A satellite has superior capabilities for mapping chl-a, TSS and CDOM (Dörnhöfer et al. 2016). The Sentinel-2 mission provides quality information for monitoring agricultural practices, land use/land cover changes, and pollution in lakes and coastal waters. Sentinel 2B images retrieved recently made it possible to study bleaching of Australia's Great Barrier Reef because of its high resolution and frequent revisits. The Sentinel-3 satellites carry modern instruments to measure oceans, land, ice, and atmosphere and can monitor aquatic biological productivity and marine pollution. Sentinel-3A was launched on February 16, 2016. Among the upcoming missions, Sentinel-6 is equipped with a radar altimeter to measure sea surface height to assist in oceanography and for climate studies.

Similarly, Landsat Data Continuity Mission (LDCM) is the next-generation Landsat satellite launched to continue the acquisition of imagery by Landsat series. This mission has Operational Land Imager (OLI) sensor that gives more accurate observations of water quality parameters, with the radiometric resolution capable of detecting chlorophyll, suspended materials, and CDOM (Lim and Choi 2015). Additionally, missions such as the Surface Water and Ocean Topography (SWOT) are expected to create a better understanding of the world's oceans and its terrestrial surface waters. Further, Environmental Mapping and Analysis Program (EnMAP), the German hyperspectral satellite has one of its missions to monitor quality of inland and coastal waters, as well as kelp (algae) mapping (Guanter et al. 2015; Foerster et al. 2016). EnMAP and other upcoming imaging spectroscopy missions, such as NASA's HyspIRI (Devred et al. 2013), Japan's Hyperspectral Imager Suite (HISUI), PRISMA, HYPXIM, and SHALOM may thus open new vistas in remote sensing-based water quality investigations (Foerster et al. 2016).

With regard to the growing interest in GIS in water quality, GPS and GIS linkages are expected to strengthen in the future, making real-time monitoring more accurate. GIS has innovative 3D displays and can overlay 3D maps over each other.

It is possible that a fourth dimension of time gets added to the capabilities that make mapping for future possibilities. This will open new avenues for predictive spatial modeling of water quality, especially in the context of potential climate change.

13.6 Conclusions

The important conclusions drawn based on the comprehensive review of applications of remote sensing and GIS in water quality monitoring and remediation are listed below:

(i) Remote sensing of water quality parameters is advantageous in terms of the spatial (synoptic scale) and temporal (past and in real time) coverage achieved and low cost that cannot be matched with efforts undertaken for field measurements and sampling.
(ii) Whether it is near real-time monitoring or analysis for past years, remote sensing is advantageous for water quality monitoring exercise. This holds true for research involving hypothesis that changes in water quality are influenced by the changes in land use/land cover pattern.
(iii) There are numerous ongoing efforts to develop atmospheric correction methods to get the best quality information on water quality parameters through remote sensing. The usefulness of atmospheric correction to reflectance data is dependent on the ecosystem being monitored. A discussion on these complex atmospheric correction algorithms is however beyond the scope of this chapter.
(iv) GIS and geostatistical techniques go hand in hand. A variety of spatial interpolation techniques such as inverse distance weighting, ordinary kriging, and universal kriging method are available for spatial analysis of water quality parameters. The choice of a method is however based on how much the resulting spatial pattern of water quality parameter obtained is accurate.
(v) Surface and groundwater vulnerability mapping for watershed/regional assessments are possible by overlaying the mapped pollutant concentrations with the land use maps. This requires processing of remote sensing satellite imagery of high resolution as well as a GIS environment that is user-friendly and supports complex spatial operations.
(vi) Use of models such as DRASTIC in conjunction with GIS have facilitated investigation of the vulnerability to groundwater contamination on a regional, rather than on a site-specific basis. This makes it possible to plan remediation of moderate to large water bodies such as lakes, long shoreline waters, and seas.
(vii) There are upcoming satellite missions aiming synoptic scale water pollution and water quality monitoring and management. In future, it is possible that the fourth dimension of time gets incorporated into GIS to address spatial modeling of water quality in the context of land use–land cover changes, natural variability of climate as well as anthropogenic climate change.

References

Abdelmalik KW (2016) Role of statistical remote sensing for Inland water quality parameters prediction. The Egyptian Journal of Remote Sensing and Space Science

Ahn HI, Chon HT (1999) Assessment of groundwater contamination using geographic information systems. Environmental Geochemistry and Health, 21(3):273–289

Akbar TA, Hassan QK, Achari G (2014) Development of remote sensing based models for surface water quality. CLEAN–Soil, Air, Water, 42(8):1044–1051

Al-Adamat RA, Foster ID, Baban SM (2003) Groundwater vulnerability and risk mapping for the Basaltic aquifer of the Azraq basin of Jordan using GIS, remote sensing and DRASTIC. Appl Geogr 23(4):303–324

Alexander RB, Smith RA, Schwarz GE, Boyer EW, Nolan JV, Brakebill JW (2007) Differences in phosphorus and nitrogen delivery to the Gulf of Mexico from the Mississippi River Basin. Environ Sci Technol 42(3):822–830

Alparslan E, Aydöner C, Tufekci V, Tüfekci H (2007) Water quality assessment at Ömerli Dam using remote sensing techniques. Environ Monit Assess 135:391

Arnold JG, Williams JR, Srinivasan R, King KW (1996) The soil and water assessment tool (SWAT) user's manual. Temple, TX

Asadi SS, Rajani G, Reddy MA (2007) Analysis and interpretation of groundwater contamination using remote sensing and GIS: a case study. Curr World Environ 2(2):115–126

Babiker IS, Mohamed MA, Terao H, Kato K, Ohta K (2004) Assessment of groundwater contamination by nitrate leaching from intensive vegetable cultivation using geographical information system. Environ Int 29(8):1009–1017

Babiker IS, Mohamed MA, Hiyama T, Kato K (2005) A GIS-based DRASTIC model for assessing aquifer vulnerability in Kakamigahara Heights, Gifu Prefecture, central Japan. Sci Total Environ 345(1):127–140

Basavarajappa HT, Manjunatha MC (2015) Groundwater quality analysis in Precambrian rocks of Chitradurga district, Karnataka, India using Geo-informatics technique. Aquat Procedia 4:1354–1365

Brekke C, Solberg AH (2005) Oil spill detection by satellite remote sensing. Remote Sens Environ 95(1):1–13

Bukata RP (2013) Retrospection and introspection on remote sensing of inland water quality: "Like Déjà Vu all over again". J Great Lakes Res 39:2–5

Chen J, Quan W (2012) Using Landsat/TM imagery to estimate nitrogen and phosphorus concentration in Taihu Lake, China. IEEE Journal of Selected Topics in Applied Earth Observations and Remote Sensing 5(1):273–280

Chen S, Fang L, Zhang L, Huang W (2009) Remote sensing of turbidity in seawater intrusion reaches of Pearl River Estuary–a case study in Modaomen water way, China. Estuar Coast Shelf Sci 82(1):119–127

Choubey VK, Subramanian V (1990) Nature of suspended solids and IRS1A-LISSI data: a case study of Tawa reservoir (Narmada Basin). Remote Sens Environ 34(3):207–215

Chu HJ, Liu CY, Wang CK (2013) Identifying the relationships between water quality and land cover changes in the tseng-wen reservoir watershed of Taiwan. Int J Environ Res Public Health 10(2):478–489

Corwin DL, Wagenet RJ (1996) Applications of GIS to the modeling of nonpoint source pollutants in the vadose zone: A conference overview. Journal of environmental quality 25(3):403–411

Corwin DL, Loague K, Ellsworth TR (1998) GIS-based modeling of non-point source pollutants in the vadose zone. J Soil Water Conserv 53(1):34–38

Curran PJ, Novo EMM (1988) The relationship between suspended sediment concentration and remotely sensed spectral radiance: a review. J Coast Res 351–368

Digenis CJ (2005) The EO-1 mission and the advanced land imager. Lincoln Laboratory Journal 15(2):161

Devred E, Turpie KR, Moses W, Klemas VV, Moisan T, Babin M, Toro-Farmer G, Forget MH, Jo YH (2013) Future retrievals of water column bio-optical properties using the Hyperspectral Infrared Imager (HyspIRI). Remote Sensing 5(12):6812–6837

Dörnhöfer K, Göritz A, Gege P, Pflug B, Oppelt N (2016) Water constituents and water depth retrieval from Sentinel-2A—A first evaluation in an Oligotrophic Lake. Remote Sens 8 (11):941

Doerffer R, Fischer J (1994) Concentrations of chlorophyll, suspended matter, and gelbstoff in case II waters derived from satellite coastal zone color scanner data with inverse modeling methods. Journal of Geophysical Research: Oceans 99(C4):7457–7466

El Saadi AM, Yousry MM, Jahin HS (2014) Statistical estimation of rosetta branch water quality using multi-spectral data. Water Sci 28(1):18–30

ESRI (1990) Understanding GIS the arc/info method. Environmental Systems Research Institute, Redlands, CA, USA

Everitt JH, Yang C, Escobar DE, Webster CF, Lonard RI, Davis MR (1999) Using remote sensing and spatial information technologies to detect and map two aquatic macrophytes. J Aquat Plant Manag 37:71–80

Fichot CG, Downing BD, Bergamaschi BA, Windham-Myers L, Marvin-DiPasquale M, Thompson DR, Gierach MM (2015) High-resolution remote sensing of water quality in the San Francisco Bay-Delta Estuary. Environ Sci Technol 50(2):573–583

Foerster S, Carrère V, Rast M, Staenz K (2016) Preface: The Environmental Mapping and Analysis Program (EnMAP) Mission: Preparing for Its Scientific Exploitation

Fraser RH, Barten PK, Pinney DA (1998) Predicting stream pathogen loading from livestock using a geographical information system-based delivery model. J Environ Qual 27(4):935–945

Frere MH, Ross JD, Lane LJ (1980). The nutrient submodel. In: Knisel W (ed) CREAMS: a field model for chemicals, runoff and erosion from agricultural management systems, vol 26. USDA Conserv Res Rep, p 340

Gholizadeh MH, Melesse AM, Reddi L (2016) Spaceborne and airborne sensors in water quality assessment. Int J Remote Sens 37(14):3143–3180

Giardino C, Candiani G, Zilioli E (2005) Detecting chlorophyll-a in Lake Garda using TOA MERIS radiances. Photogramm Eng Remote Sens 71(9):1045–1051

Giupponi C, Vladimirova I (2006) Ag-PIE: a GIS-based screening model for assessing agricultural pressures and impacts on water quality on a European scale. Sci Total Environ 359(1):57–75

Griffith JA (2002) Geographic techniques and recent applications of remote sensing to landscape-water quality studies. Water Air Soil Pollut 138(1):181–197

Griffith JA, Martinko EA, Whistler JL, Price KP (2002) Interrelationships among landscapes, NDVI, and stream water quality in the US Central Plains. Ecol Appl 12(6):1702–1718

Guanter L, Kaufmann H, Segl K, Foerster S, Rogass C, Chabrillat S, Straif C (2015) The EnMAP spaceborne imaging spectroscopy mission for earth observation. Remote Sens 7(7):8830–8857

Harding LW, Itsweire EC, Esaias WE (1995) Algorithm development for recovering chlorophyll concentrations in the Chesapeake Bay using aircraft remote sensing, 1989–91. Photogramm Eng Remote Sens 61(2):177–185

Hellweger FL, Schlosser P, Lall U, Weissel JK (2004) Use of satellite imagery for water quality studies in New York Harbor. Estuar Coast Shelf Sci 61(3):437–448

Hestir EL, Khanna S, Andrew ME, Santos MJ, Viers JH, Greenberg JA, Ustin SL (2008) Identification of invasive vegetation using hyperspectral remote sensing in the California Delta ecosystem. Remote Sens Environ 112(11):4034–4047

Holme AM, Burnside DG, Mitchell AA (1987) The development of a system for monitoring trend in range condition in the arid shrublands of Western Australia. Rangel J 9(1):14–20

Huan H, Wang J, Teng Y (2012) Assessment and validation of groundwater vulnerability to nitrate based on a modified DRASTIC model: a case study in Jilin City of northeast China. Sci Total Environ 440:14–23

Hunter PD, Tyler AN, Carvalho L, Codd GA, Maberly SC (2010) Hyperspectral remote sensing of cyanobacterial pigments as indicators for cell populations and toxins in Eutrophic Lakes. Remote Sens Environ 114(11):2705–2718

Jensen JR, Hodgson ME, Garcia-Quijano M, Im J, Tullis JA (2009) A remote sensing and GIS-assisted spatial decision support system for hazardous waste site monitoring. Photogramm Eng Remote Sens 75(2):169–177

Jerlov NG (1976) Marine optics. Elsevier, Amsterdam, p 194

Jha MN, Levy J, Gao Y (2008) Advances in remote sensing for oil spill disaster management: state-of-the-art sensors technology for oil spill surveillance. Sensors 8(1):236–255

Jones KB, Riitters KH, Wickham JD, Tankersley RD Jr, O'Neill RV, Chaloud DJ, Neale AC (1997) An ecological assessment of the United States mid-Atlantic region: a landscape atlas. United States Environmental Protection Agency. Office of Research and Development, Washington DC, 20460. EPA/600/R-97/130

Jupp DL, Kirk JT, Harris GP (1994) Detection, identification and mapping of cyanobacteria—using remote sensing to measure the optical quality of turbid inland waters. Mar Freshw Res 45 (5):801–828

Kaba E, Philpot W, Steenhuis T (2014) Evaluating suitability of MODIS-Terra images for reproducing historic sediment concentrations in water bodies: Lake Tana, Ethiopia. Int J Appl Earth Obs Geoinf 26:286–297

Kallio K, Koponen S, Ylöstalo P, Kervinen M, Pyhälahti T, Attila J (2015) Validation of MERIS spectral inversion processors using reflectance, IOP and water quality measurements in boreal lakes. Remote Sens Environ 157:147–157

Kazakis N, Voudouris KS (2015) Groundwater vulnerability and pollution risk assessment of porous aquifers to nitrate: modifying the DRASTIC method using quantitative parameters. J Hydrol 525:13–25

Keith DJ, Schaeffer BA, Lunetta RS, Gould RW Jr, Rocha K, Cobb DJ (2014) Remote sensing of selected water-quality indicators with the hyperspectral imager for the coastal ocean (HICO) sensor. Int J Remote Sens 35(9):2927–2962

Kelly JR, Harwell MA (1990) Indicators of ecosystem recovery. Environ Manag 14(5):527–545

Kilham NE, Roberts D, Singer MB (2012) Remote sensing of suspended sediment concentration during turbid flood conditions on the Feather River, California—a modeling approach. Water Resour Res 48:1

Kirk JTO (1983) Light and photosynthesis in aquatic ecosystems. Cambridge University Press, Cambridge, p 509

Knox RC, Sabatini DA, Canter LW (1993) Subsurface transport and fate processes. Lewis Publishers

Kulkarni A (2011) Water quality retrieval from Landsat TM imagery. Procedia Comput Sci 6:475–480

Kumar DN, Reshmidevi TV (2013) Remote sensing applications in water resources. J Indian Inst Sci 93(2):163–188

Kutser T (2012) The possibility of using the Landsat image archive for monitoring long time trends in coloured dissolved organic matter concentration in lake waters. Remote Sens Environ 123:334–338

Leifer I, Lehr WJ, Simecek-Beatty D, Bradley E, Clark R, Dennison P, Hu Y, Matheson S, Jones CE, Holt B, Reif M (2012) State of the art satellite and airborne marine oil spill remote sensing: application to the BP Deepwater Horizon oil spill. Remote Sens Environ 124:185–209

Lillesand TM, Kiefer RW (1994) Remote sensing and image interpretation, 3rd edn. John Wiley and Sons, New York

Lim J, Choi M (2015) Assessment of water quality based on Landsat 8 operational land imager associated with human activities in Korea. Environ Monit Assess 187(6):384

Liu Z, Li Y, Li Z (2009) Surface water quality and land use in Wisconsin, USA—a GIS approach. J Integr Environ Sci 6(1):69–89

Maidment DR (1996) GIS and hydrologic models of non-point source pollution in subsurface water. In: Applications of GIS to the Modeling of Non-Point Source Pollutants in the Vadose Zone, pp 163–174

Markogianni V, Dimitriou E, Karaouzas I (2014) Water quality monitoring and assessment of an urban Mediterranean lake facilitated by remote sensing applications. Environ Monit Assess 186 (8):5009–5026

Martins VS, Barbosa CCF, de Carvalho LAS, Jorge DSF, Lobo FDL, Novo EMLDM (2017) Assessment of Atmospheric Correction Methods for Sentinel-2 MSI Images Applied to Amazon Floodplain Lakes. Remote Sensing 9(4):322

Matthews MW, Bernard S, Winter K (2010) Remote sensing of cyanobacteria-dominant algal blooms and water quality parameters in Zeekoevlei, a small hypertrophic lake, using MERIS. Remote Sens Environ 114(9):2070–2087

Mueller TG, Pusuluri NB, Mathias KK, Cornelius PL, Barnhisel RI, Shearer SA (2004) Map quality for ordinary kriging and inverse distance weighted interpolation. Soil Sci Soc Am J 68 (6):2042–2047

Munn RE (1988) The design of integrated monitoring systems to provide early indications of environmental/ecological changes. Environ Monit Assess 11(3):203–217

Nas B, Berktay A (2010) Groundwater quality mapping in urban groundwater using GIS. Environ Monit Assess 160(1):215–227

NASA (2010) NASA images show oil's invasion along Louisiana coast. National Aeronautics and Space Administration, https://www.nasa.gov/topics/earth/features/oil20100602.html. Accessed on 8 Sept 2017

Navulur KCS, Engel BA (1998) Groundwater vulnerability assessment to non-point source nitrate pollution on a regional scale using GIS. Trans ASAE 41(6):1671–1678

Nelson SA, Soranno PA, Cheruvelil KS, Batzli SA, Skole DL (2003) Regional assessment of lake water clarity using satellite remote sensing. Journal of Limnology 62(1s):27–32

NOC (2009) National Oceanography Centre. Areas of research in the HERMIONE project, funded under the European Commission's Framework Seven Programme, Southampton, UK

Odermatt D, Gitelson A, Brando VE, Schaepman M (2012) Review of constituent retrieval in optically deep and complex waters from satellite imagery. Remote Sens Environ 118:116–126

Oikonomidis D, Dimogianni S, Kazakis N, Voudouris K (2015) A GIS/remote sensing-based methodology for groundwater potentiality assessment in Tirnavos area, Greece. J Hydrol 525:197–208

Ouyang D, Bartholic J, Selegean J (2005) Assessing sediment loading from agricultural croplands in the Great Lakes Basin. J Am Sci 1(2):14–21

Palmer SC, Kutser T, Hunter PD (2015) Remote sensing of inland waters: challenges, progress and future directions. Remote Sens Environ 157:1–8

Petrovic M, Eljarrat E, Lopez de Alda MJ, Barceló D (2004) Endocrine disrupting compounds and other emerging contaminants in the environment: a survey on new monitoring strategies and occurrence data. Anal Bioanal Chem 378:549. https://doi.org/10.1007/s00216-003-2184-7

Ritchie JC, Cooper CM (2001) Remote sensing techniques for determining water quality: applications to TMDLs. In: TMDL Science Issues Conference, Water Environment Federation, Alexandria, VA, Mar 2001, pp 367–374

Ritchie JC, Zimba PV, Everitt JH (2003) Remote sensing techniques to assess water quality. Photogramm Eng Remote Sens 69(6):695–704

Ritchie JC, McHenry JR, Schiebe FR, Wilson RB (1974) The relationship of reflected solar radiation and the concentration of sediment in the surface water of reservoirs, Remote Sensing of Earth Resources Vol. III (F. Shahrokhi, editor), The University of Tennessee Space Institute, Tullahoma, Tennessee, pp 57–72

Robertson DM, Schwarz GE, Saad DA, Alexander RB (2009) Incorporating uncertainty into the ranking of SPARROW model nutrient yields from Mississippi/Atchafalaya River basin watersheds1. JAWRA J Am Water Resour Assoc 45(2):534–549

Secunda S, Collin ML, Melloul AJ (1998) Groundwater vulnerability assessment using a composite model combining DRASTIC with extensive agricultural land use in Israel's Sharon Region. J Environ Manage 54(1):39–57

Shaffer MJ, Hall MD, Wylie BK, Wagner DG (1996) NLEAP/GIS approach for identifying and mitigating regional nitrate-nitrogen leaching. In: Applications of GIS to the Modeling of Non-Point Source Pollutants in the Vadose Zone, pp 283–294

Shen L, Xu H, Guo X (2012) Satellite remote sensing of harmful algal blooms (HABs) and a potential synthesized framework. Sensors 12(6):7778–7803

Simis SG, Peters SW, Gons HJ (2005) Remote sensing of the cyanobacterial pigment phycocyanin in turbid inland water. Limnol Oceanogr 50(1):237–245

Singh CK, Shashtri S, Mukherjee S, Kumari R, Avatar R, Singh A, Singh RP (2011) Application of GWQI to assess effect of land use change on groundwater quality in lower Shiwaliks of Punjab: remote sensing and GIS based approach. Water Resour Manag 25(7):1881–1898

Singh A, Jakubowski AR, Chidister I, Townsend PA (2013) A MODIS approach to predicting stream water quality in Wisconsin. Remote Sens Environ 128:74–86

Sinkevich MG Jr, Walter MT, Lembo AJ, Richards BK, Peranginangin N, Aburime SA, Steenhuis TS (2005) A GIS-based ground water contamination risk assessment tool for Pesticides. Groundw Monit Remediat 25(4):82–91

Sridhar BM, Vincent RK, Witter JD, Spongberg AL (2009) Mapping the total phosphorus concentration of biosolid amended surface soils using LANDSAT TM data. Science of the total environment 407(8):2894–2899

Srinivasa Rao Y, Jugran DK (2003) Delineation of groundwater potential zones and zones of groundwater quality suitable for domestic purposes using remote sensing and GIS. Hydrol Sci J 48(5):821–833

Srivastava PK, Singh SK, Gupta M, Thakur JK, Mukherjee S (2013) Modeling impact of land use change trajectories on groundwater quality using remote sensing and GIS. Environ Eng Manag J (EEMJ) 12:12

Tassan S (1993) An improved in-water algorithm for the determination of chlorophyll and suspended sediment concentration from Thematic Mapper data in coastal waters. Int J Remote Sens 14(6):1221–1229

Tebbs EJ, Remedios JJ, Harper DM (2013) Remote sensing of chlorophyll-a as a measure of cyanobacterial biomass in Lake Bogoria, a hypertrophic, saline–alkaline, flamingo lake, using Landsat ETM+. Remote Sens Environ 135:92–106

Thirumalaivasan D, Karmegam M, Venugopal K (2003) AHP-DRASTIC: software for specific aquifer vulnerability assessment using DRASTIC model and GIS. Environ Model Softw 18 (7):645–656

Tiktak A, van der Linden A, Leine I (1996) Application of GIS to the modeling of pesticide leaching on a regional scale in the Netherlands. In: Applications of GIS to the modeling of non-point source pollutants in the vadose zone, pp 259–281

Toming K, Kutser T, Laas A, Sepp M, Paavel B, Nõges T (2016) First experiences in mapping lake water quality parameters with Sentinel-2 MSI imagery. Remote Sens 8(8):640

USGS (2016) At least one pharmaceutical chemical was detected in all 59 streams sampled. Department of the Interior, U.S. Geological Survey, Reston, VA. Accessed on 1 Sept 2017. Available at https://www.usgs.gov/news/pharmaceuticals-commonly-detected-small-streams-southeastern-united-states

Volpe V, Silvestri S, Marani M (2011) Remote sensing retrieval of suspended sediment concentration in shallow waters. Remote Sens Environ 115(1):44–54

Yenilmez F, Kuter N, Emil MK, Aksoy A (2011) Evaluation of pollution levels at an abandoned coal mine site in Turkey with the aid of GIS. Int J Coal Geol 86(1):12–19

Yüzügüllü O, Aksoy A (2011) Determination of Secchi Disc depths in Lake Eymir using remotely sensed data. Procedia-Soc Behav Sci 19:586–592

Zhang Y, Pulliainen JT, Koponen SS, Hallikainen MT (2003) Water quality retrievals from combined Landsat TM data and ERS-2 SAR data in the Gulf of Finland. IEEE Trans Geosci Remote Sens 41(3):622–629

Zhang Y, Zhang Y, Shi K, Zha Y, Zhou Y, Liu M (2016) A Landsat 8 OLI-Based, Semianalytical Model for Estimating the Total Suspended Matter Concentration in the Slightly Turbid Xin'anjiang Reservoir (China). IEEE J Sel Top Appl Earth Obs Remote Sens 9(1):398–413

MIX
Papier aus verantwortungsvollen Quellen
Paper from responsible sources
FSC® C105338

If you have any concerns about our products,
you can contact us on
ProductSafety@springernature.com

In case Publisher is established outside the EU,
the EU authorized representative is:
Springer Nature Customer Service Center GmbH
Europaplatz 3, 69115 Heidelberg, Germany

Printed by Libri Plureos GmbH
in Hamburg, Germany